T0141870

Studies in Computational Intelligence

Volume 735

Series editor

Janusz Kacprzyk, Polish Academy of Sciences, Warsaw, Poland
e-mail: kacprzyk@ibspan.waw.pl

About this Series

The series "Studies in Computational Intelligence" (SCI) publishes new developments and advances in the various areas of computational intelligence—quickly and with a high quality. The intent is to cover the theory, applications, and design methods of computational intelligence, as embedded in the fields of engineering, computer science, physics and life sciences, as well as the methodologies behind them. The series contains monographs, lecture notes and edited volumes in computational intelligence spanning the areas of neural networks, connectionist systems, genetic algorithms, evolutionary computation, artificial intelligence, cellular automata, self-organizing systems, soft computing, fuzzy systems, and hybrid intelligent systems. Of particular value to both the contributors and the readership are the short publication timeframe and the worldwide distribution, which enable both wide and rapid dissemination of research output.

More information about this series at http://www.springer.com/series/7092

Krzysztof Dyczkowski

Intelligent Medical Decision Support System Based on Imperfect Information

The Case of Ovarian Tumor Diagnosis

 Springer

Krzysztof Dyczkowski
Faculty of Mathematics and Computer
    Science, Department of Imprecise
    Information Processing Methods
Adam Mickiewicz University
Poznań
Poland

ISSN 1860-949X   ISSN 1860-9503   (electronic)
Studies in Computational Intelligence
ISBN 978-3-319-88363-2   ISBN 978-3-319-67005-8   (eBook)
https://doi.org/10.1007/978-3-319-67005-8

Printed on acid-free paper

This Springer imprint is published by Springer Nature
The registered company is Springer International Publishing AG
The registered company address is: Gewerbestrasse 11, 6330 Cham, Switzerland

I dedicate this book to my Family: my wife Anna, sons Kuba and Michał, and to my Parents

Preface

This monograph is the result of scientific research conducted between 2012 and 2016 by an interdisciplinary team of scientists from the Department of Imprecise Information Processing Methods at Faculty of Mathematics and Computer Science of Adam Mickiewicz University (AMU) in Poznań and The Division of Gynecologic Surgery of the Poznań University of Medical Sciences (PUMS). The AMU team includes prof. dr hab. Maciej Wygralak, dr inż. Anna Stachowiak, dr Patryk Żywica, and mgr Andrzej Wójtowicz. The PUMS team is composed of: prof. dr hab. n.med Dariusz Szpurek, dr hab. n.med. Rafał Moszyński, and dr n.med. Sebastian Szubert. The author of this monograph is the project leader.

Medical diagnosis and the proper differentiation of ovarian tumors are very important issues. Every year, about 2500 women die in Poland due to this type of cancer and over 14,000 in the USA. Poland is at the top of the list with the highest mortality due to this cancer. In addition, these tumors are difficult to diagnose because they give too few unambiguous symptoms in the early stage. Differentiating this form of tumor is in many cases difficult and requires a physician with extensive experience, advanced medical equipment, and repeated use of imprecise and incomplete data.

Physicians specializing in this type of disease have noted the need for a computerized system to support the diagnosis of such tumors, which would help the less-experienced physicians in the diagnosis and differentiation of tumors. There is also a need for a central database for standardized gathering of medical information on cases of patients treated in different centers in Poland, which would allow us to improve the developed diagnostic algorithms in the future.

The book outlines the theoretical foundations and the description of the *OvaExpert* system created by our team. The system supports the physician in the diagnosis of ovarian tumors using computational intelligence methods with particular focus on the problem of data incompleteness. The first part of the book introduces basic information on medical diagnosis and the diagnosis of ovarian tumors. The statistics concerning this disease have also been presented. In the next part of the book, we have introduced the necessary elements of fuzzy set theory and its extensions such as IVFSs and IFSs. The key part of the monograph is the description of original algorithms based on the cardinalities of IVFSs. The algorithm descriptions are supplemented with the analysis of the efficacy of the presented algorithms, performed on the data of patients treated in the Poznań medical center. The last chapter describes the *OvaExpert* system, its key elements, and technologies and analyzes the prediction efficacy of algorithms used in the system against other existing methods. The book ends with a brief description of the ongoing and planned research and development works connected with the system.

I would like to thank the whole team involved in the project, without whom this book could not have been written. In particular, I would like to thank prof. dr hab. Maciej Wygralak for supporting me throughout the years, for pointing me in scientific directions and motivating to do further research. I extent my gratitude to colleagues from my Department: dr Anna Stachowiak, dr Andrzej Wójtowicz, and dr Patryk Żywica for their daily cooperation and hard work, commitment, as well as conceptual and technical support. I am grateful to the doctors from the Poznan University of Medical Sciences who cooperated with us, and especially to dr hab. Rafał Moszyński and dr n.med. Sebastian Szubert whom I would like to thank for introducing me to the medical realm of knowledge and for tremendous support during the creation of the system, and to prof. dr hab. Dariusz Szpurek, for his interest in the subject and creating favorable conditions for cooperation between our teams.

Finally, I would like to thank my beloved wife, Anna, for her understanding throughout the whole time which I devoted to my research and for her unwavering support in difficult times.

Poznań, Poland
April 2017

Krzysztof Dyczkowski

Contents

Acronyms and Abbreviations

Acc	Accuracy
ADNEX	The Assesment of Different Neoplasis in the adneXa
Alc	Alcazar's model
AUC	Area under an ROC curve
BI-RADS	Breast Imaging Reporting Data System
BMI	Body Mass Index
CBR	Case-Based Reasoning
CEE	Central and Eastern Europe
COG	Center of Gravity
DBMS	Database management system
Dec	Decisiveness
EBM	Evidence-Based Medicine
FCS	Family of all Crisp Sets
FDA	Food and Drug Administration
FE	FECount
FFS	Family of all Fuzzy Sets
FG	FGCount
FIGO	International Federation of Gynecology and Obstetrics
FL	FLCount
FN	False Negative
FP	False Positive
FSC	*Ovaexpert* decision method based on counting
GDM	Group Decision Making
GFE	Family of all generalized cardinal numbers
GI-RADS	Gynecolologic Imaging Reporting and Data System
GP	General Practitioner
IFS	Atanassov's Intuitionistic Fuzzy Set
IOTA	International Ovarian Tumor Analysis Group
IVFC	OvaExpert decision method based on Interval-Valued Fuzzy Classifier
IVFS	Interval-Valued Fuzzy Set

LR	Logistic Regression
LR1	IOTA Logistic regression-based model 1
LR2	IOTA Logistic regression-based model 2
Max	Maximum
Min	Minimum
MSK	Memorial Sloan-Kettering Cancer Center
MVC	Model-View-Controler
NICE	National Institute for Health and Care Excellence
OEA	OvaExpert decision method based on aggregation
OECD	The Organization for Economic Cooperation and Development
ORM	Object-Relational Mapping
OWA	Ordered Weighted Averaging
PACS	Picture Archiving and Communication System
PBD	Percent Breast Density
PDI	Polytomous Discrimination Index
PPV	Positive Predictive Value
Prec	Precision
RIS	Radiology Information System
RMI	Risk of Malignancy Index
ROC	Receiver Operating Characteristic
ROMA	Risk of Ovarian Malignancy Algorithm
SaaS	Software as a Service
SD	Doppler Index
SDR	Standardized Death Rate
Sen	Sensitivity
SM	Sonomorphological Index
Sneg	Family of all strong negations
Spec	Specificity
SR	IOTA Simple Rules diagnostic model
Supp	Support of a fuzzy set
Tim	Timmerman's logistic regression model
TN	True Negative
TNR	True Negative Rate
TP	True Positive
TPR	True Positive Rate
USG	Ultrasonography

List of Figures

List of Tables

Abstract

This book deals with a computer-supported medical diagnosis with particular focus on ovarian tumor diagnosis. The book presents theoretical foundations (both medical and mathematical) of the intelligent *OvaExpert* system which supports the decision-making process in tumor diagnosis. The main purpose for creating *OvaExpert* was to support a gynecologist in the process of predicting the malignancy of ovarian tumors by applying the existing diagnostic models and using modern methods of computational intelligence which allow for imprecision and imperfection of the medical data, both of which are common features of the everyday medical practice. We had good reasons to focus on this particular cancer. Ovarian cancer is difficult to diagnose and has high mortality, especially in the Central and Eastern Europe. The book presents novel methods based on interval-valued fuzzy sets and the theory of their cardinalities. The algorithms applied have been verified by analyzing their efficacy on real data.

Chapter 1
Introduction

Computers have long been applied in medicine. Information systems are used for medical data storage, hospital and clinic management and for supporting diagnosis and treatment. First support systems for decision-making in medicine appeared at the beginning of the 1970s (see De Dombal et al. [25]). The increasing computational capacity of computers and major developments in machine learning techniques resulted in their enhanced efficacy and made them more applicable in medical diagnosis. These system are highly effective in solving many diagnostic problems. This is true especially for common diseases for which there is access to large number of cases.

The situation is less satisfactory for diseases which are less common and thus the access to large number of well-depicted cases is limited. Lack of centralized system for gathering uniform data from many medical institutions is also a problem. If such databases exist they are gathered in a specific medical center and are not accessible to others. Another problem is lack of access to full required diagnostics (e.g. due to unavailability of proper diagnostic machines or high cost of diagnostic examinations), which contributes to ambiguities and omissions in patient's record. In addition, by their very nature, medical descriptions are often imprecise and ambiguous. In most cases, they are descriptive and terminology used in them is not standardized. Their quality often depends on the education of the doctor (it also depends on the center where he or she was educated) as well as the doctor's experience.

The existing situation calls for the use of unconventional data modeling and reasoning methods. It requires methods factoring in both the imprecision and incompleteness of the data. Those methods must ensure high efficacy for disease entities for which there are no sufficiently large databases available.

This book focuses on the data collection and medical diagnosis stage. The selection and process of treatment is a different and broad topic requiring separate research and analysis.

The main purpose of this book is the description of the intelligent system OvaExpert which supports the physician in diagnosing ovarian tumor. This type of tumor is not that common compared to, for example, breast cancer, but it does have a high mortality rate (see World Health Organization [145]). Detection rate for this disease is still very low. This is because in many cases (about 50%) it is very difficult to determine the presence of neoplastic features during a routine gynecological

© Springer International Publishing AG 2018
K. Dyczkowski, *Intelligent Medical Decision Support System Based on Imperfect Information*, Studies in Computational Intelligence 735, https://doi.org/10.1007/978-3-319-67005-8_1

examination (see Szpurek [119]). This is why the development of methods facilitating effective diagnosis of this type of tumor is so important. Equally important is differentiation of its form, since depending on its malignancy and correct diagnosis, it is possible to implement the right treatment, which may result in better prognosis for recovery in the future and maintaining good quality of life. The problem here might be both diagnosing a patient with a malignant tumor as healthy, which may result in non-treatment, or diagnosing the disease in a healthy patient which may lead to implementing a risky and exhaustive therapy, diminished comfort of living and psychological damage.

Those reasons were the rationale behind creating the diagnostic system OvaExpert which would support less-experienced physicians in making the right diagnostic decisions. An important element of the system is the central platform for collecting medical data. This will facilitate the improvement of predictive methods and support of physicians through the whole process of diagnosing a patient (e.g. by indicating the next diagnostic steps, potentially raising the quality of the diagnosis). The methods developed when creating the system may also be used in the future, not only for ovarian tumor diagnosis, but also other diseases in which there is a need to differentiate a form of a given symptom; and there are many diagnostic models available with possible deficiencies in data occurring at the same time.

The book presents both the theoretical foundations of the system construction as well as detailed description of the system itself which is furnished not only with diagnostic functions but also with such essential elements as data anonymization and security. Results of experiments confirming the efficacy of the prepared and implemented methods will be shown in which special emphasis was put on factoring in gaps in the data and their imprecision that is key element in every doctor's practice.

Uncertainty is a serious problem in everyday medical practice and it is observed and described in medical literature. As noted in Han et al. [51] there are many meanings and types of imprecision in medicine with each of them having a different effect on the diagnosis. The imprecision types may be divided into objective (caused by the complexity or nature of the phenomenon) or subjective (caused by a personal opinion or doctor's interpretation) or caused by the low quality of the information, e.g. due to gaps in the data. Imperfection is meant in the sense that the values of some entities, variables, etc. are imprecisely known, which can be modeled and processed by using fuzzy sets theory, but some uncertainty is also assumed which is meant to be related to the degrees of membership in the respective fuzzy sets and this two aspects are jointly modeled by using the concept of an interval-valued fuzzy set.

The OvaExpert system introduced a completely novel approach to the imprecision connected with data imperfection (see Dyczkowski et al. [36], Wójtowicz [152]). The aim of the system is to store and process uncertain data in such a way as to be able to obtain information essential to make an effective diagnosis while also indicating the uncertainty level with which the information is suggested.

The book is structured as follows. Chapter 2 presents basic information concerning medical diagnosis. A general approach to the construction of the decision-making support systems is described. The chapter also deals with the concept of tumor and its basic types. Statistical data on ovarian cancer in Poland and the world are presented as well. Basic terms connected with the problems of ovarian tumor diagnosis are introduced as well as the existing differentiation models of ovarian tumor forms and, finally, methods of evaluating the quality of decisions made by such predictive models.

Chapter 3 introduces the theoretical foundations of algorithms construction used in the system. Special emphasis has been given to the theory of fuzzy sets, which are the basis for further algorithms, with particular focus on the cardinality theory of such objects. Basic terms on data aggregation are also introduced.

Chapter 4 is devoted to the issue of cardinality of interval-valued fuzzy sets. The concepts of interval-valued fuzzy sets (IVFSs) and Atanassov's intuitionistic fuzzy sets (IFSs) are introduced. Decision algorithms based on the IVFS cardinalities are presented along with methods of their optimization, results obtained by them and evaluation of their efficacy in differentiating ovarian tumors. The chapter also deals with source data sets used in their evaluation.

Chapter 5 depicts the basics of the structure and implementation of the OvaExpert system. The chapter also contains the description of the technologies used in the developed system architecture. User interface is depicted with particular focus on bipolar presentation of the diagnosis. Basic diagnostic modules are described along with the results of their decision-making efficacy on test data.

The book ends with the summary of the obtained results. It also includes plans for future development of the OvaExpert system and computational methods used therein.

Chapter 2
Medical Foundations

It is curious that most medical curricula do not explicitly teach the student of medicine the meanings of the words 'diagnosis' and 'disease'. The student is assumed to form his or her own mental picture of what is understood by these terms. A result of this is that most doctors have a largely intuitive idea of the meaning of these concepts, as they find out when they start thinking about them.

van Herk [54]

In this chapter we present the basics of medical decision support systems, indicating their types and elements they should include. We also introduce the basic information on ovarian tumors as well as existing models supporting their differentiation. The techniques used to evaluate the efficacy of prediction methods are also discussed. We point to some problems connected with medical data imperfection.

2.1 Computer-Based Systems for Decision-Making Support in Medicine

In recent years the use of computer-based systems in medicine has increased significantly. They are applied not only in data gathering (organizational support systems) but also in supporting doctors in diagnosis and treatment of patients. Creating such systems is possible thanks to the development of intelligent decision-making systems which can have different structures depending on their application and computational methodology employed in them. As seen in Belle et al. [11] they can be divided as follows:

- expert systems, rule-based and case-based systems—where the inference is based on the collected medical case data or based on a rule base provided by experts,

© Springer International Publishing AG 2018
K. Dyczkowski, *Intelligent Medical Decision Support System Based on Imperfect Information*, Studies in Computational Intelligence 735, https://doi.org/10.1007/978-3-319-67005-8_2

- signal processing systems—medical data from a variety of sources (i.e. ECG) are analyzed to identify important features that facilitate medical decision making,
- machine learning—data-driven systems that can predict and classify medical cases.

The current systems frequently use the synergy of several of the above models at the same time.

The key role of a medical information system in terms of diagnosis and treatment is to support the decision-making process. A physician in everyday practice must constantly make decisions, for example, when choosing a diagnostic scheme or selecting the right medicine. Today doctors have access to a great number of data about each patient, but this huge amount of data can sometimes cause difficulty in making a decision. According to the requirements of Evidence Based Medicine (EBM) methodology (see Evidence-Based Medicine Working Group [39]), a doctor should make clinical decisions based on the best available research. This is possible through the use of appropriate computer systems (see Tadeusiewicz [125]).

In medical diagnosis support systems, we can distinguish two categories depending on the degree and the methods of using medical databases and knowledge bases. Database systems are primarily based on comparing the currently diagnosed patient with clinically verified patient descriptions in the database, which is called case-based reasoning (CBR) (see Aamodt and Plaza [1]). These methods use statistical tools and machine learning techniques. Systems with knowledge bases are mainly based on the experience of expert doctors, represented in the form of rules, often fuzzy (see Oniko et al. [90]).

Clinical decision making usually involves checking and analyzing large amounts of data. The amount of data gathered and processed, combined with the growing speed of their acquisition, makes it impossible to collect data in a traditional way—in patient records. An important part of the process of diagnosis and treatment is the cost, which should also be taken into account. According to the postulate of evidence-based medicine (EBM), decision-making should be based on verified facts and latest research results. Due to a large amount of clinical data collected, and the ever increasing demands for quality and efficacy of decisions, there is a greater interest in clinical decision support systems and their increased application in clinical practice. When analyzing the concept of clinical decision support system, it may be claimed that the concept is not precisely defined. You can find the definition of this system as a computer system, helping clinicians in any way to make clinical decisions. This definition is very broad and usually covers the following classes of clinical decision support systems (after Slowinski et al. [124]):

- Information management systems. Systems of this type provide medical personnel with access to knowledge and medical data. Depending on the type of information being shared and processed, they are divided into systems for managing medical knowledge (e.g. electronic versions of lexicons or atlases) and patient management systems. The last mentioned are often referred to as electronic health record.
- Alerts and reminders systems. Systems of this type are used in laboratories for reporting abnormal results, in hospital pharmacies for checking and indicating adverse interactions between prescribed drugs, in clinics and hospital departments

for reminding clinical staff of specific tasks (e.g. vaccinations) and warning of possible risks (e.g. patient allergies to certain medicines).
• Suggestion systems. Systems of this type provide suggestions for decisions concerning individual patients, factoring in their condition. They use a wide variety of techniques to generate hints and decision-making suggestions, from the implementation of the adopted procedures and schemes to the artificial intelligence techniques.

When considering the classification of clinical decision support systems, it turns out, it does not include systems for supporting group work and communication which exist as an independent class of telemedical systems. Telemedical systems have been present in clinical practice for as long as information management systems have, but it was the development of broadband networks, which made them more commonplace. With faster and easier exchange of knowledge and experience between medical staff, these systems strongly support clinical decision making and should therefore be included in the above classification as an additional category of decision support systems. The previously cited definition of clinical decision support systems, treating them as computer systems helping the medical staff to make decisions in any way, is very general. Based on this definition, any computer system that is used in clinical settings can be considered a decision support system. Therefore, it is worth considering a more precise variant definition limited to computer systems which, based on information describing the current condition of the patient and built-in clinical knowledge, provide evaluations, warnings, or patient-specific recommendations. Thus, the definition includes suggestion systems as well as alarms and reminders systems (see Slowinski et al. [124]).

Clinical decision support systems are helpful in solving various types of clinical problems and are used in different places and by different users. Despite such diversity, however, all of them should meet the following requirements (after Slowinski et al. [124]):

• they should adapt to the existing working methods and procedures of the medical staff—this also means that they should be available at a time and place where a decision problem requiring support occurs,
• they should be integrated with information management systems (especially electronic health record) to share and exchange information,
• they should provide reliable and well-substantiated hints and suggestions, with the final decision always made by the medical staff.

According to Slowinski et al. [124], systems that will force users to change their fixed course of action (for example, they will require a physician to change the standard order of tests or leave the patient to use the system) will not be accepted in practice, even if they offer enhanced functionality. Ideally, a clinical decision support system should be tailored to the accepted and recognized procedures and should be available at the time and place of the decision problem. Decision support systems should be integrated with data collection and management systems so as not to force the doctor to re-enter previously collected data. An important element of a modern system is that it works on mobile devices such as a tablet or smartphone and thus can be used outside the doctor's office and without any infrastructure. Another key feature is its user-friendliness and intuitive use. It should be noted that in the clinical decision-making process the final decision is always taken by the physician. That is why such systems will always be used only to advise and assist the physician, and they will not determine the final diagnoses.

2.1.1   Solutions in Medical Diagnosis Support in the World

Worldwide, there are many systems supporting medical centers in their organizational functioning and supporting the medical personnel in diagnosis and treatment.[1] Only a few examples of such systems will be presented here, showing their relevance and variety of applications.

One such system is the ZynxEvidence system which is manufactured by Zynx-Health (see ZynxHealth [161]). It offers rule-based clinical decision support modules based on medical facts. By integration with medical databases, the system generates detailed care plans based on the patient's condition.

Another example may be a system offered by the Thomson Reuters Micromedex department. It provides a point-of-care CDS Suite. It is used in over 3,500 hospitals in 83 countries. It provides support for ensuring safety in drug prescribing, disease management, toxicology, etc. The system offers mobile applications that provide information about medicines and their interactions (see Micromedex Solutions [80]).

In the countries such as the United Kingdom and the United States, the AxSys Healthcare Management System is very popular (see AxSys Technology Ltd [7]). This package consists of such modules as: medical database systems, disease knowledge systems (including tumors), patient flow management, and telemedicine solutions.

Another example from the UK is the Arezzo Pathways system (see Elsevier [58]) produced by InferMed (taken over in 2015 by Elsevier). It offers support for clinical decision making and the designation and optimization of clinical pathways. It reduces the cost of health care and improves the quality of service. Currently, it includes 41 thematic areas containing 1,000 clinical symptoms or patient treatment scenarios.

The second category of systems includes the specialized systems for supporting the diagnosis. The UK has the Prodigy system (see Clarity Informatics [17]). It is accredited by the National Institute for Health and Care Excellence (NICE). It was created in collaboration with scientists and Clarity Informatics company. The system specializes in general medicine. It is able to propose a treatment scenario for a given medical case. It currently contains 341 thematic areas consisting of 1,000 clinical symptoms or treatment scenarios.

---

[1]In this book we used an analysis report prepared for the purpose of the OvaExpert project (see Trąpczński [135]).

In many countries the Gideon system is currently used (see GIDEON [44]). In addition to the epidemiological or therapeutic modules, it contains a diagnostic module. It facilitates diagnostic predictions and optimal diagnostic schemes. This system requires high quality and complete information to make a diagnosis. A similar solution created in the US by the Laboratory of Computer Science at the Massachusetts General Hospital is the DXplain system (see Laboratory of Computer Science at the Massachusetts General Hospital [73]). It supports the diagnosis of various medical problems (24,000 diseases including their symptoms, etiology and pathology). This system presents a list of possible diagnoses for a given patient based on a modified Bayesian inference, including the strength of the links between the symptoms and the disease and the frequency of their occurrence, the severity and frequency of the disease, the descriptions of the individual diseases, and references to the relevant publications—hence it can be used as a textbook. This system has been in development since 1987 and is available to registered users (doctors, medical students) as an application in the Saas model.

2.1.2 Solutions in Cancer Diagnosis Support in the World

The use of computer imaging systems supporting cancer diagnosis has a long tradition worldwide. The first computerized diagnostic support system, which was approved in 1998 by the FDA (Food and Drug Administration), was ImageChecker by R2 Technology, which was based on scanned mammography picture, and able to find clusters of microcalcifications and tumors (see Przelaskowski [107]). Another systems were SecondLook by CADx Medical Systems (2002), MammoReader by iCAD (2002), KODAK Mammography Computer-Aided Detection System proposed by Carestream Health for analogous mammography (2004). Among other commercial systems approved for clinical application we can mention: B-CAD by Medipattern for sonomammography (2005, breast cancer diagnosis), IQQA-Chest by EDDA Technologies/Philips (2004, chest x-ray; detection and measurement of tumors), xLNA by Philips Medical Systems (detects tumors with a diameter of 5 mm), Rapid Screen CAD by Riverain Medical (the first CAD system for chest x-ray approved by the FDA in 2001), Lung VCAR by GE Healthcare; Syngo Lung CAD by Siemens Medical Solutions (CT scan of the lungs), CAD-Lung by Median Technologies, or IQQA-Liver by EDDA (CT of the liver) (see Przelaskowski [107]).

Tumor diagnostic support systems can also be discussed in the context of teleradiology, particularly in the United States, where companies such as USARAD Holdings, Inc operate (see USARAD Holdings, Inc. [136]). The company's operation is based on a SaaS model that allows physicians in smaller locations to send pictures (including mammography or computed tomography) to radiologists working in the USARAD network via the Radiology Information System (RIS), which uses the technology of the Picture Archiving and Communication System (PACS) (see USARAD Holdings, Inc. [137]). As a result, GPs who have requested a radiological interpretation can receive an answer within an average time of 15 min. Another example is iCAD Inc (see iCAD Inc. [57]) which offers solutions in the field of computer-image diagnosis support. IReveal system allows automating the mammogram analysis process by providing information such as percent breast density (PBD), area of dense tissue, total breast area etc.

Information systems supporting the diagnosis of cancer are also used in the field of digital and contact thermography. Eidam Diagnostics Corporation (see Eidam Diagnostics Corporation [37]), selling contact thermography equipment around the world allows medical practitioners to use their devices to send their thermograms to a central database. There, the results are compared to up to 1.5 million existing thermographic data. On this basis, within a few seconds, the doctor receives a report of the results and information about possible irregularities.

Computer systems are also used in non-imaging diagnostic methods. One of the better known systems is the Watson system created by IBM (see IBM Corporation [56]). The Watson system facilitates the analysis of a wide variety of both structured and unstructured data. Based on the data, it can draw conclusions. This is a general purpose system that answers questions asked in a natural language, enables advanced text analysis (context-based meaning detection), and adapts the question to the available resources. Thanks to its analytical capabilities, it performs well in medical diagnosis, including oncology. In 2013, based on the agreement between IBM and Memorial Sloan-Kettering Cancer Center (MSK) in New York, and a private health care provider Wellpoint, Watson was introduced into commercial application (as Watson for Oncology) for hospitals needing expertise in the field of oncology (see Steadman [115]). The system analyzes the patient's medical records (also recorded in a natural language) and provides recommendations for further diagnostic tests and treatment (according to the EBM) methodology. The decision-making system uses a large body of data, which includes the MSK database of publications and justifications as well as over 290 medical journals, over 200 books and 12 million pages of text.

As in the case of systems for supporting the broadly-defined medical diagnosis, there are also many noteworthy systems for supporting decision-making in cancer diagnosis. The Lisa (Leukemia Intervention Scheduling and Advice) program in the United Kingdom has been developed since 2002 in pediatric oncology. Created by the Information Systems Development team at the British Cancer Research UK, the Children's Cancer Group (CCG) and the Advanced Computation Laboratory (ACL), the system used an engine developed by the Advanced Computation Laboratory and Cancer Research UK, which is also the basis for many commercial applications already in use or in the clinical trial phase (see Bury et al. [13, 14], OpenClinical [92]). With regard to early detection of cancer, the ERA system, piloted by the UK NHS Information Authority, was an important step. The ERA system was integrated with the EMIS Practice Management System prior to implementation in two health care facilities in Leicester and Southampton. The system was to be used by 10 general practitioners and 60 GPs (see OpenClinical [91]).

As indicated by Belle et al. [11], when developing new information systems supporting cancer diagnosis, more emphasis should be put on integrating knowledge from different sources, i.e. molecular tests or imaging. They also note the need to create better schemes for assessing the quality and effectiveness of existing systems in this field. This is the trend the OvaExpert system described in this book fits in (see also: OvaExpert Project Homepage [95]).

2.2 Ovarian Tumors

2.2.1 The Concept of a Tumor

Tumor (neoplasm) is an uncontrollable and abnormal division of cells. The causes of neoplasms are not clear, but they are related to mutations in the genes of the proteins that make up the tumor. They may be affected by various physical or chemical factors, viral infections, diet, and use of stimulants. The neoplastic process is multi-stage and can progress slowly (e.g. several years) or aggressively fast (e.g. several weeks) (see National Cancer Institute [88]).

The most important issue in the pathology of tumors is their division into benign and malignant types. Benign tumors develop slowly and are limited to the original site and do not metastasize. Usually, they can be removed surgically and therefore, in most cases, are not life-threatening conditions. Malignant tumors are a diverse group, both morphologically (type of cells) and clinically (course of the disease). Their key feature, unlike non-malignant tumors, is the formation of local and remote metastases and often micro-metastases which are not detectable at the time of the first diagnosis, but which are the source of dissemination and failure of treatment. The time and the extent of their detection is one of the factors determining the degree of malignancy (see Cooper [18], National Cancer Institute [88]).

The largest group of malignant tumors are those originating from the epithelial cells (e.g. from the ovarian epithelium), that is carcinomas (cancers) (see Kosary [70]). In the initial stage cancer is usually a restricted tumor, and as the disease progresses, it can invade the surrounding healthy tissues and, subsequently, ulcerative, often bleeding dissemination foci occur. The inherent trait of malignant neoplasms, including carcinomas, is their local and distant metastasizing due to the penetration of cancer cells into the lymphatic system and blood. The rate of growth and histopathological differentiation are very important for prognosis.

Although the differences between the two types of tumors are quite significant, the final distinguishing can be made only on postoperative histopathological examination. The key problem in the case of ovarian tumors is preoperative differentiation between malignant and benign ovarian tumors, which determines the method of treatment. However, appropriate preoperative diagnosis is not always possible, and sometimes is very difficult to make (see Maśliński and Ryżewski [76]).

In the case of ovarian tumors, there is a third type of tumor—borderline malignant tumors. These tumors have features of both malignant and benign tumors. They represent about 15% of all ovarian tumors. In most cases, borderline tumors are detected relatively early and usually have a good prognosis (see Morotti et al. [84]).

The staging of ovarian tumor is determined based on the FIGO scale (International Federation of Gynecology and Obstetrics) upon histopathological testing (of postoperative or biopsied samples) (see Heintz et al. [53]). The current classification (guidelines from 2014 are available at Society of Gynecologic Oncology [112]) of the FIGO clinical staging for ovarian tumors is presented in Table 2.1.

2.2.2   The Scale of the Problem

Diagnosis and effective treatment of cancers are among the greatest challenges of modern medicine. Despite increasingly better methods of detecting and treating cancer, it is still one of the most common causes of death (see Eurostat [38]).

Ovarian cancer is the seventh most common cancer in women, and in the whole population it is on the 18th place, with 239,000 new cases diagnosed in 2012 (after World Cancer Research Fund International [144]). In terms of prognosis, ovarian cancer is the worst gynecological cancer, with the lowest 5-year survival rate (see Polish Society of Oncology [101]). This type of cancer is so dangerous that in the early stages it often does not show any symptoms, so it is not uncommon to diagnose it when it already reached an advanced stage (see Polish Society of Oncology [101]). Despite intensive research into the etiology of ovarian cancer, the cause of this cancer is still unknown. There are many theories on this subject, such as that the causes of the disease may be genetic errors or increased levels of hormones before and during ovulation, which stimulates abnormal cell growth. However, the exact causes remain unexplained (see Szubert [122], Ovarian Cancer National Alliance [96]).

According to the data from the American National Cancer Institute (NCI) (see National Cancer Institute [87]), in the United States alone, the estimated number of new cases of ovarian cancer for 2016 is slightly over 22,000 and the number of deaths is over 14,000, representing 2.4% of all deaths due to cancer (see Fig. 2.1). The same data show that 60% of all diagnosed cases are at an advanced stage, i.e. the cancer has spread to more organs. Another 19% of cases are ovarian cancer, which has spread to the adjacent lymph nodes, which can also cause it to spread to other organs. When interpreting this information, it is important to note the survival of the patient depending on the moment of the detection of cancer. Thus, in the case of early detection (i.e. localized tumor stage), the 5-year relative survival rate is 92.1%, for the regional stage 73.2%, but only 28.3% for the advanced stage.

Malignant ovarian tumors constitute a serious problem in Poland. Compared to the European Union countries, Poland has high incidence and still low cure rates. At the same time, in recent years, significant progress has been made in the molecular diagnosis of the genetic markers for this disease (see Polish Society of Oncology [100]). Among malignant tumors of females, ovarian cancer was the second leading cause of illness (after endometrial cancer) and the fourth leading cause of death among all malignancies in women population in Poland in 2012 (see Polish Society of Oncology [101]). Ovarian cancer occurs in about 3,500 women each year, and the number of deaths is about 2,500. The prognosis of patients diagnosed with malignant ovarian

Table 2.1 FIGO classification for ovarian tumors (after Society of Gynecologic Oncology [112])

STAGE I: Tumor confined to ovaries			
IA	Tumor limited to 1 ovary, capsule intact, no tumor on surface, negative washings		
IB	Tumor involves both ovaries otherwise like IA.		
IC	Tumor limited to 1 or both ovaries		
	IC1	Surgical spill	
	IC2	Capsule rupture before surgery or tumor on ovarian surface	
	IC3	Malignant cells in the ascites or peritoneal washings.	
STAGE II: Tumor involves 1 or both ovaries with pelvic extension (below the pelvic brim) or primary peritoneal cancer			
IIA	Extension and/or implant on uterus and/or Fallopian tubes		
IIB	Extension to other pelvic intraperitoneal tissues		
STAGE III: Tumor involves 1 or both ovaries with cytologically or histologically confirmed spread to the perito			
IIIA	Positive retroperitoneal lymph nodes and /or microscopic metastasis beyond the pelvis		
	IIIA1	Positive retroperitoneal lymph nodes only	
		IIIA1(i) Metastasis $\leq$ 10 mm	
		IIIA1(ii) Metastasis > 10 mm	
	IIIA2	Microscopic, extrapelvic (above the brim) peritoneal involvement $\pm$ positive retroperitoneal lymph nodes	
IIIB	Macroscopic, extrapelvic, peritoneal metastasis $\leq$ 2 cm $\pm$ positive retroperitoneal lymph nodes. Includes extension to capsule of liver/spleen		
IIIC	Macroscopic, extrapelvic, peritoneal metastasis > 2 cm $\pm$ positive retroperitoneal lymph nodes. Includes extension to capsule of liver/spleen.		
STAGE IV: Distant metastasis excluding peritoneal metastasis			
IVA	Pleural effusion with positive cytology		
IVB	Hepatic and/or splenic parenchymal metastasis, metastasis to extraabdominal organs (including inguinal lymph nodes and lymph nodes outside of the abdominal cavity)		

tumors is significantly worse in Poland than in most European countries, which is a consequence—first of all—of late detection and inaccurate (against the guidelines) course of action, which in some cases takes place in centers with insufficient means and unskilled medical staff (see Polish Society of Oncology [100]).

The OECD data point to the still increasing incidence of malignancy among men and women in most of the analyzed countries (see OECD.Stat [89]). Comparing results for Poland with the values for 10 countries with the highest incidence per 100 000 men or women in 2012, it can be observed that the incidence in Poland is declining. By narrowing the focus to ovarian cancer, however, Poland is still in the top range in terms of both the incidence and mortality associated with ovarian cancer (after Ferlay et al. [41]): 18.1 in 100,000 women for incidence and 10.3 in 100,000 women for mortality. Poland is ranked 4th in the world in terms of incidence of ovarian cancer (see Table 2.2).

In Poland, ovarian cancer is the fifth major disease among neoplastic diseases in women, after breast, lung, colorectal, and uterine corpus cancer, and before cervical cancer (after Ferlay et al. [41]). This is attributed to the regional tendencies associated with this disease. And so, in the CEE region the incidence of ovarian cancer in 2012 was twice as high than in East Asia (after World Cancer Research Fund International [144]). Annually, about 2,500 women die of ovarian cancer in Poland. According to national data, the value of this indicator fluctuated between 6.8 and 7 per 100,000 in the past six years. According to GLOBOCAN estimates, the standardized death rate (SDR) for ovarian cancer in Poland in 2012 was 7.3 per 100,000 and was among the highest in Europe (see Polish Society of Oncology [101]).

Risk factors in ovarian tumors are usually unpredictable and mostly difficult to eliminate. The only available primary prevention strategies are based on the early identification of women at high risk for ovarian cancer due to hereditary reasons, including BRCA1 and/or BRCA2 mutation carriers, and taking special care of them, including, inter alia, risk-reducing surgical procedures for ovarian cancer (see Polish Society of Oncology [101]).

Table 2.2 Countries with the highest incidence of ovarian cancer in 2012 (*source* World Cancer Research Fund International [144])

Rank	Country	Age-standardized rate per 100 000 (worldwide).
1	Fiji	14.9
2	Latvia	14.2
3	Bulgaria	14.0
4	**Poland**	**13.6**
5	Serbia	12.8

2.2.3   Diagnosis and Differentiation of Tumors

Proper diagnosis and treatment of adnexal masses is a big challenge for modern medicine. This applies to both malignant tumors (e.g. ovarian cancer) as well as non-malignant ones, and also non-neoplastic tumors, e.g. functional or inflammatory lesions. From a practical point of view, the important question is whether an ovarian tumor identified in a patient is malignant, benign or non-neoplastic because each of these tumors requires different treatment. For ovarian malignancies, an abdominal surgery is required and performed in a center specializing in the treatment of malignant ovarian tumors. On the other hand, benign tumors should be operated using the least invasive methods (i.e. laparoscopic surgery) performed in regional hospitals. Non-neoplastic tumors, in most cases, are functional lesions that do not require surgery because they usually subside spontaneously after the follow-up period. Unfortunately, at present, there is no method for effective preoperative diagnosis and tumor characterization is performed during surgery using histopathological examination (microscopic analysis of the tumor sample). The available diagnostic methods only establish the probability of a specific tumor occurrence.

An effective and fast diagnosis is required in order to choose the best therapeutic method while at the same time protecting the patient from unnecessary surgery. It not only saves lives, but also ensures patient's better mental health and reduces the costs of treatment. For example, a proper selection of type and place for conducting surgical procedures has a significant impact on the future of the women being treated (see du Bois et al. [12]). The key factors here are the experience and qualifications of the physician who diagnoses the patient and performs the surgery. It is extremely important to avoid unnecessary surgical procedures due to tumors not suspected of malignancy. The physician should adhere to the principle that says doctors should set bounds to surgical interventions so as to do at least what is possible and at most what is necessary (see Moszyński [85]).

The development of imaging methods, including mainly ultrasonography, significantly increased the quality of assessment of the type of malignant lesions. This facilitates preliminary diagnosis of tumor malignancy, its size and location. It also allows us to evaluate how extensive the cancer process is. Ultrasound tests performed by an experienced physician are highly sensitive and specific, also in the

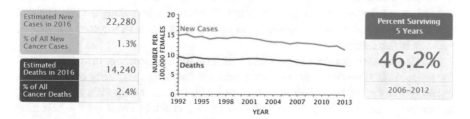

Fig. 2.1 Ovarian cancer—number of new cases and deaths per 100,000 people (*source* National Cancer Institute [87])

case of specific diagnoses (see Moszyński [85], Valentin [138]). The main problems with the ultrasonography of the ovaries are their small size and mobility, which hinders localization and standardization of the exam. An important improvement in cancer diagnosis is the use of Doppler techniques. They generate a qualitative description of tumor vasculature (color—blood flow) and a quantitative one as well (pulsation—flow velocity, resistance).

Unfortunately, there is a group of tumors (so-called diagnostically difficult) whose classification is difficult even for an experienced physician using the best equipment (see Moszyński [85], Valentin et al. [139]). Therefore new diagnostic methods based on, for example, biochemical or Doppler scales are sought. It is because of the existence of this type of tumors that many mathematical models have been developed—to evaluate the character of tumor. The existing models are described in Sect. 2.2.4. There are also second-degree tests which use a single diagnostic test and the subsequent tests are applied depending on the outcome. Every stage that follows should improve the quality of the diagnosis (see Moszyński [85]).

It should be noted here that we are constantly using the concept of quality of diagnosis. In medicine, and especially in cancer diagnosis, it is quite natural that diagnoses are made with greater or lesser uncertainty. It cannot be completely eliminated, thus one should be aware of it. A very comprehensive study on the uncertainties in medicine can be found in the book by Hatch [52]. The author, a practicing physician, and at the same time knowledgeable about statistics, argues that a good doctor is one who not only realizes the existence of imprecision but accepts it without trying to ignore it.

As mentioned earlier, the main problem in the proper diagnosis of ovarian cancer is the lack of clear symptoms in the early stages of the disease. Very often they are non-specific and often associated with gastrointestinal ailments (see Moszyński [85]). Nevertheless, it is the collection of such data by a general practitioner and then their analysis (e.g. frequency and duration), which could influence the early diagnosis of the disease (see Moszyński [85], Goff et al. [45]). In this situation, most diagnoses are made at an advanced stage of the disease. Unfortunately, only in the early detection of this potentially fatal disease can one hope for an improvement in the treatment outcome (see Moszyński [85]).

Therefore, it is very important to create a support system for the gynecologist in tumor diagnosis. Such a system would help a GP to collect patient data and assist in deciding whether to continue the diagnostic process or to refer a patient to a specialized center. The system would also support specialists in making decisions about tumor character and would facilitate data gathering in order to improve diagnostic methods in the future.

Proper diagnosis (and this is possible with a modern system supporting the physician) would also reduce the cost of treatment and improve the quality of life for the patient (e.g. by reducing the stress associated with incorrect diagnosis and avoiding unnecessary surgery). Proper diagnosis also has a crucial impact on the selection of proper treatment (e.g. laparoscopic or classic surgery). Time is also an important factor, because some malignant tumors require specialized surgical treatment during the first surgery (see Szpurek [119]).

Ovarian cancer is a complex and multi-faceted disease. The number of factors that allow a physician to make a differential assessment is very high. Those factors include past pregnancies, contraception, the occurrence of cancer in the family, menopausal status (see Moszyński et al. [86]). Despite numerous attempts, a universal list of factors has not been established to help decide if a tumor is malignant or not. Additional problems the physician might face are the limited amount of time devoted to the patient and the inability to perform certain specialized medical tests.

IOTA (International Ovarian Tumor Analysis Group) is a major contributor to the standardization of data collection and creation of new models. This international group of highly experienced researchers from the best academic centers has created a unified ultrasound assessment of adnexal tumors. Through their application, IOTA collected a large number of medical cases, based on which diagnostic scales were developed. The IOTA data description methods are currently a standard in the description of ultrasound examination (see Timmerman et al. [133]). They allow relatively easy and precise communication between researchers dealing with ovarian tumors (see Moszyński [85]). One of the elements of examination of tumor suggested by IOTA is a subjective assessment of the nature of its malignancy. It consists in determining whether a particular tumor is *definitely non-malignant/malignant*, *probably non-malignant/malignant*, *uncertain*. A subjective assessment by an experienced clinician using appropriate ultrasound equipment is currently considered the most effective method for predicting the malignancy of adnexal tumors (see Moszyński [85], Timmerman et al. [130]).

In oncological diagnosis, a considerable hope lies in finding various substances in the serum of diagnosed patients, the so-called markers. They can help in early detection (screening) and in the differentiation of detected tumors. In the case of ovarian cancer, CA125 marker concentration is important. Unfortunately, it shows some limitations, especially in early cancer stage. Another promising marker is HE4. It appears to be more sensitive in early cancer, and more resistant to false-positive results in non-malignant neoplasms. So far, none of them is sufficiently precise to be considered an effective screening tool (see Moszyński [85]).

A very important criterion in the differentiation of ovarian tumors is medical history. It includes such elements as the patient's age, BMI, menopausal status, history of births. Although not related to the tumor itself, medical history often provides key information necessary for the tumor differentiation process. As research has shown, one of the most important factors for the efficacy of diagnostic scales is the patient's menopausal status (see Moszyński et al. [86]). Another important factor is the patient's age. For example, in young patients, the level of CA125 markers in the blood can fluctuate.

2.2.4   Diagnostic Models

With the development of diagnostic methods based on ultrasonography, blood markers and clinical trials, the medical community has noted the need for clinicians to prepare decision-making methods to support the differentiation of tumors. A significant incentive to create such models was an observation that the combination of several parameters achieves very high diagnostic efficacy in particular cases.

Most of the methods currently under development are based on a multidimensional statistical analysis (mainly logistic regression), but also on machine learning methods, e.g. neural networks. When creating this type of model, it is very important to have access to sufficiently large numbers of medical cases. Very often such data have to be collected over several years. Hence, such models are created only in the best medical centers specializing in gynecological oncology. It is important to point out the very important role of IOTA bringing together the best medical centers which exchange data. The article by Stukan et al. [117] presents a review of the existing decision indexes. The authors analyzed 65 different medical studies on this subject from 1990 to 2011 and compared more than 20 diagnostic models, describing their efficacy. Some of them have been made available on their website as interactive calculators (see Stukan [116]).

Below, we present the most important diagnostic models used in modern diagnosis of ovarian cancer.

2.2.4.1   ADNEX

ADNEX (The Assesment of Different Neoplasis in the adneXa) is the newest model created in 2014 by the IOTA group (see Van Calster et al. [141]). It was created as a result of many years of research conducted in 24 scientific centers from 10 countries, which gathered data of almost 6,000 clinical cases.

The scale includes three clinical parameters: age, CA125 level and type of center (it was included because the authors noted that the probability that a patient treated in an oncological center has a malignant disease is greater). It also includes 6 ultrasound examination features, such as the largest tumor dimensions or the number of papillary projections. The developed model is able to distinguish and calculate the probability of five different types of ovarian tumors, including differentiation of benign and malignant lesions (see Van Calster et al. [141]). The structure of the model is based on the multinomial logistic regression and polytomous discrimination index (PDI) (see Van Calster et al. [140]). The calculator computing the model has been made available by the IOTA group as a web and mobile application (see IOTA [62]). The high efficacy of this model has been confirmed on external data (e.g. see Szubert et al. [123]).

2.2.4.2   Alcazar's Model

This is a scoring system developed in 2003 by a team of Spanish scientists led by J.L. Alcazar from the University of Pamplona. The model was based on the data from over 600 women. It is based on four features of the ultrasound examination (presence of papillary projections, presence of solid elements, localization of blood vessels and blood flow parameters in Doppler). Each feature is scored appropriately and the final value is between 0 and 12 points. The tumor is considered malignant if the result is greater than 6 (see Alcazar et al. [2]).

2.2.4.3   GI-RADS

The GI-RADS model (Gyneclologic Imaging Reporting and Data System) was created in 2009 by researchers from two medical centers in Spain (Pamplona and Madrid) led by J.L. Alcazar (see Alcazar et al. [2].) It was created mainly to facilitate communication between ultrasonographers and clinicians. When creating this system, the researchers drew from the existing BI-RADS system (Breast Imaging Reporting Data System), which was created in the American College of Radiology in 1993, and its forth version is approved by the FDA. Halls [50] GI-RADS is a very simple five-point classification based on the evaluation of the ovary ultrasonogram. In prospective studies, the authors have confirmed the high efficacy of this system (see Amor et al. [3]). Despite these good results, the usefulness of this system is questioned by some experts because it is based on a subjective judgment of the physician, which requires wide experience of the physician performing the ultrasound examination.

2.2.4.4   IOTA LR1 and LR2

These are models by the IOTA group based mainly on the ultrasonographic examination, Doppler techniques and clinical data. Their creation contributed to the widespread use of standards for the description of these exams. LR1 and LR2 models were created based on the data from over 1,000 patients from 9 research institutions in 5 countries (see Timmerman et al. [132, 134]). Both models were created using the logistic regression (hence the name LR).

The first model (LR1) includes 12 clinical data from a patient (such as age, hormone therapy), and features from ultrasonography (such as vascularization, the largest tumor dimension). The second model (LR2) requires fewer tests. Six tests were selected with the highest significance for differentiation: age, presence of ascites, presence of blood flow within a papillary projection, largest diameter of the solid component, presence of irregular internal wall, and presence of acoustic shadows. Both models estimate the probability of a malignant change in a patient. They assume that the tumor is malignant if the result is greater than 0.1 (see Timmerman et al. [132]). Both models exhibit very high sensitivity and specificity, including in external validation (see Stukan et al. [117]). Their great advantage (especially LR2) is their simplicity and the fact that they do not require biochemical tests. The disadvantage is the requirement that the description of ultrasound examinations be done by an experienced diagnostician, familiar with the description methodology.

2.2.4.5   IOTA Simple Rules (SR)

This is a scale published in 2008 by the IOTA group and based on the ultrasonographic examination and Doppler methods. It was created by analyzing over 1.2 thousand cases of ovarian tumors. As a result of the analysis, ten simple rules were developed. Five of them point in the direction of malignancy of the tumor (M-rules) and five rules indicate the direction of the benign tumor (B-rules). The advantage of this method is its simplicity. It is also very effective in predicting a large number of tumor types. According to the authors, it reaches 93% sensitivity and 90% specificity (see Timmerman et al. [131]). The problem of the model is, unfortunately, a high percentage of non-diagnostic tumors, presenting both malignant and benign features. The percentage is around 25%. The rules and examples are presented in Fig. 2.2.

The IOTA Group makes available on its website an easy-to-use smartphone application and a spreadsheet with which the user, upon entering the patient's data, receives the prediction of the tumor character. The authors point out that the model was developed based on patients with adnexal tumors selected for surgery because surgical intervention was needed to properly verify the histological diagnosis. This means that the received formulas and rules have not been developed and tested for women with adnexal cysts and tumors that have not been classified for surgical treatment. As a result, the model cannot be used for conservatively treated adnexal tumors. Especially in patients with simple cysts, the risk of cancer can be overestimated. The authors also point out that the model cannot replace training and experience in ultrasonography and cannot compensate for ultrasound equipment of poor quality (see IOTA [61]).

2.2.4.6   Risk of Malignacy Index (RMI)

The RMI scale (Risk of Malignacy Index) was created in 1990 as a result of research done by scientists from the Department of Obstetrics and Gynaecology at the London Hospital (see Jacobs et al. [63], Davics et al. [23]). This method combines the measurement of CA125 marker with the results of ultrasound and the menopausal status of the patient. The ultrasonography describes the following features: the presence of multilocular cyst, the presence of solid components, the presence of ascites, the presence of intra-abdominal metastases, and the bilateral changes. The result of the ultrasound scan is scored on a scale of 0, 1 or 3 (1 for one change and 3 for more changes). For menopausal status the value of this parameter is 1 if the patient is premenopausal, and 3 if the patient is postmenopausal. The result is the product of these three parameters. It is assumed that the cutoff point is 200, that is a score above this value indicates malignancy.

The advantage of this model is its availability because it does not require a complex set of markers and advanced ultrasound examinations. The drawback is that the model shows high efficacy only in postmenopausal patients (see Stukan et al. [117]).

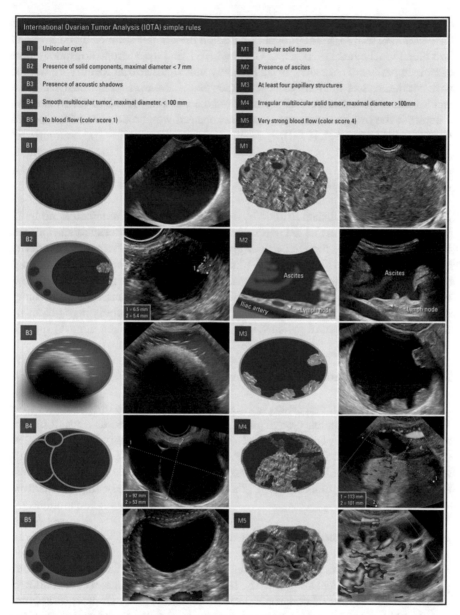

Fig. 2.2 IOTA simple rules (*source* IOTA: Educational material [60])

2.2.4.7 ROMA

This is a popular diagnostic method. The name ROMA comes from the first letters of Risk of Ovarian Malignancy Algorithm. The method was created in 2009 by a team of researchers from several leading medical research centers in the United States (see Moore et al. [83]). The model is based on the combination of blood levels of two blood biochemical markers, CA125 and HE4, and the menopausal status of the examined patient. This model is significantly more sensitive in identifying women with ovarian cancer than the RMI scale (see Sect. 2.2.4.6). Its important feature is lack of the subjective aspect of the ultrasound evaluation. Just like the RMI scale, it achieves the best performance in postmenopausal women because of the CA125 marker (see Stukan et al. [117]).

2.2.4.8   Doppler Index (SD)

The SD score was published in 2004 by a group of clinicians from Poznań led by D. Szpurek (see Szpurek et al. [121]). It was created as a result of data analysis of almost 500 patients treated at the Division of Gynecologic Surgery, Poznan University of Medical Sciences. It is based on the ultrasonographic Doppler scale. The scale is based on 5 selected features (number of vessels, their location and layout, blood flow velocity and presence of protodiastolic notch in the arterial vessels), and additionally, it factors in the menopausal status. This model exhibits particularly high predictive efficacy in postmenopausal patients.

2.2.4.9   Sonomorphological Index (SM)

The SM scale is the second scale published in 2005 by a group of researchers from Poznań led by D. Szpurek (see Szpurek [119]). It was created as a result of data analysis of almost 700 patients. It is based on seven features mainly from transvaginal ultrasonography. Each feature is scored on a scale from 0 to 4 points, and the result is the sum of all points. The tumor is assessed as malignant if the scale value exceeds 8. When tested on a group of patients, the model has proved to be highly effective.

2.2.4.10   Timmerman's Logistic Regression Model

This is a model developed by a team led by D. Timmerman from Louven in Belgium (see Timmerman et al. [129]. The model uses, as input data, a combination of ultrasnographic results of transvaginal ultrasonography, using Doppler techniques, and CA125 antigen level. It is based on logistic regression. Based on the results of 191 patients between 18 and 93 y.o., the model is highly sensitive with still quite high specificity. The study described in Timmerman et al. [129] showed that the combination of the results of CA125 and relevant ultrasonography parameters may yield very good discriminative results.

2.3 Methods of Classification Evaluation

A very important aspect of construction and application of tumor malignancy classification (prediction) methods is to evaluate their efficacy (prediction quality). In this book we focus on binary classification, in which we divide the decision into two classes: positive (malignant tumors, which also include borderline malignant neoplasms which require the same treatment as ovarian cancer), and negative (benign tumors and non-neoplastic changes). In addition, we allow a situation in which a classifier may not make a decision due to the data being of too low quality (lack of large portion of the information).

In medicine, many quality classification measures are applied. Below we present the most commonly used measures of prediction quality.

2.3.1 Types of Classification Errors

In order to construct metrics that measure the quality of a binary classifier, the first step is to identify the types of errors that it can make. As mentioned above, the classifier indicates, on the basis of the input data, that the object belongs to two classes: Positive (P) and Negative (N) (see Japkowicz and Shah [64]). We assume that we have the input data available (most often symptoms of the disease) as well as the actual final diagnosis (e.g. after histopathology) for a certain group of patients. By comparing the binary classifier results on the input data with the actual diagnoses we can distinguish four cases (shown in the Table 2.3):

- TP—True Positive (hit)—specifies the number of positive class objects that have been classified as positive;
- TN—True Negative (correct rejection)—specifies the number of negative class objects that have been classified as negative;
- FP—False Positive (Type I error)—specifies the number of negative class objects that have been classified as positive;
- FN—False Negative (Type II error)—specifies the number of positive class objects that have been classified as negative.

Table 2.3 Types of classification errors

Actual	Predicted	
	P' (Malignant)	N' (Benign)
P (Malignant)	TP (correct positive decision)	FN (type II error: false negative)
N (Benign)	FP (type I error: false positive)	TN (correct negative decision)

2.3.2 Performance Measures

Based on the error types defined in Sect. 2.3.1 measures can be defined to assess the quality of classification in various aspects. Below we present the most common quality measures.

2.3.2.1 Accuracy

Accuracy (Acc.)—this is a measure of the ratio of the number of correctly classified objects to all evaluated objects. In other words, this is the probability of a proper classification:

$$Acc = \frac{TP + TN}{TP + TN + FP + FN}.$$ (2.1)

This is the most intuitive measure, but despite its simplicity it is not always the best measure. Especially in the case of unbalanced number of positive and negative cases in test sample (see Japkowicz and Shah [64]). Additionally, accuracy does not work in situations where it is more important for us to have no errors of a given type.

2.3.2.2 Sensitivity

Sensitivity (Sen.) (other terms for this measure: true positive rate—TPR, hit rate, recall) specifies how many of the positive class objects are properly categorized. It can be interpreted as the probability that the classification will be correct, provided that the case is positive, i.e. the probability that the test performed for a cancer patient will show that the tumor is malignant. We define this measure as follows:

$$Sens = \frac{TP}{P} = \frac{TP}{TP + FN}.$$ (2.2)

2.3.2.3 Specificity

Specificity (Spec.) (also: true negative rate—TNR) shows how often the model correctly classifies objects from the negative class. In other words, it is the probability that the classification will be correct, provided that the case was negative, i.e. the probability that a person with a benign tumor will show that the tumor is benign. We define this measure as follows:

$$Spec = \frac{FP}{N} = \frac{FP}{FP + TN}.$$ (2.3)

It is worth noting that sensitivity decreases with the increase in the classification errors of positive objects whereas specificity decreases with the increase of the classification errors of negative objects. By using both of these measures, we are able to find a model that correctly classifies the largest number of objects, regardless of class. Additionally, if one of the error types is more undesirable, we may pay more attention to one of them, for example, if the incorrect classification of a sick person has worse outcomes than the incorrect classification of a healthy person.

2.3.2.4 Precision

Positive precision (Prec.) (also positive predictive value—PPV) is the ratio of the number of positive results to the number of all results classified as positive. It answers the question: If the test result is positive, what is the probability that a person has a malignant tumor:

$$Prec = \frac{TP}{TP + FP}.$$ (2.4)

2.3.2.5 F-Measure

Measures such as sensitivity (2.2) and specificity (2.3) should be considered simultaneously in pairs to assess the quality of classification. However, there are measures that allow us to evaluate a classifier using a single value (an example of such a measure is the previously introduced accuracy (see Sect. 2.3.2.1) which, however, has some disadvantages, described earlier).

The most commonly used measure for the balance between precision and sensitivity is the *F-measure*, which is their harmonic mean (this measure does not factor in true negative results).

$$F_\alpha = (1 + \alpha^2) \frac{Prec \cdot Sens}{\alpha^2 \cdot Prec + Sens}.$$ (2.5)

Parameter α allows us to adjust the weight that we apply to precision.

The most commonly used measure is its special case for parameter $\alpha = 1$, in this case precision and sensitivity are equally important:

$$F_1 = 2 \frac{Prec \cdot Sens}{Prec + Sens} = \frac{2TP}{2TP + FP + FN}.$$ (2.6)

2.3.2.6 ROC Curves and AUC

The performance measures presented in previous sections concerned binary classifiers. In medicine, we often use classifiers giving a continuous output (e.g. naive Bayesian classifiers) which gives the final decision result only after setting an appro-

Fig. 2.3 Example of ROC
curve for a classifier based on
CA125 and HE4 biomarkers,
for which $AUC = 0.927$

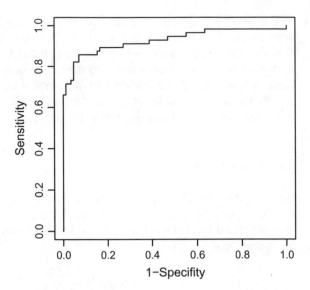

priate threshold value (the so-called cut-off point). In this situation, we need a method allowing a selection of the optimal threshold and evaluation of the resulting classifier. This we can do with the ROC curves and their analysis (see Fawcett [40], Powers [102]).

ROC curves (Receiver Operating Characteristic) provide a combined description of the sensitivity and specificity of the classifier. This method is widely used in the assessment of classification in medical systems. ROC curves are also useful for comparing different classifiers.

ROC curve is a two-dimensional graphical plot where on the X axis we have 1-specificity values (False positive rate) and on the Y axis sensitivity (True positive rate). Such a graph allows us to present the relationship between sensitivity and specificity. The result of the discrete classifier is represented as a point in the ROC space. Point $(0, 1)$ represents the perfect classification. In the case of a classifier that returns a continuous output, different cutoff levels are generated, thus obtaining a discrete classifier. In this way, if the output of the classifier is above the cutoff point, we accept a positive decision, and a negative decision if the classifier is below the cutoff point. Thus, for each cutoff point we receive a point in the ROC plot space (different sensitivity and specificity). A wider analysis of the properties of ROC curves can be found in Fawcett [40].

An example of the ROC curve for a simple classifier based on two CA125 and HE4 biomarkers is shown in Fig. 2.3.

AUC (Area under an ROC curve). In many situations (e.g. when comparing classifiers) we would like to receive a single value describing the expected quality of classification. This is what AUC offers. AUC is the area of the unit graph under the ROC curve and takes values in the [0, 1] range. It can be interpreted as the probability that a classifier will assign a higher rank to a randomly selected positive or negative case (assuming a positive result has a higher rank than a negative one). In this sense, the concept is closely related to the Wilcoxon statistics (see Fawcett [40]).

2.3.2.7 Decisiveness

For classifiers that operate on poor quality data (e.g. incomplete data), in some applications, it is necessary to consider a situation in which the classifier has insufficient information to make a sufficiently certain decision. This is often the case in medical applications when insufficiently certain decision can have serious consequences for the patient. In this case, it would be better for the decision system to report inability to make a decision and suggest additional diagnostic tests. In the case of such classifiers, additional classes of errors N0 and N1 are introduced, describing the number of non-decisions in the positive and negative case, respectively (Table 2.4).

Decisiveness measure determines in how many cases the classifier was able to make a decision:

$$Dec = \frac{TP + TN + TP + FN}{TP + TN + FP + FN + N0 + N1}. \tag{2.7}$$

It should be noted that in the case of classifiers that permit the option of not making a decision, the earlier measures, i.e. sensitivity or specificity, are calculated only for those cases in which the classifier has generated a decision. Hence, in this case, all calculated measures for a given classifier should always be evaluated with reference to the decisiveness measure.

2.3.2.8 Cost Matrix

In many applications (often including medical ones) the above measures do not reflect the actual required quality of the classifier. This is the case when the significance of the individual classes of errors (actual effects of wrong decisions) are different. For example, in the medical diagnosis of ovarian tumors the situation when the system diagnoses a tumor as benign and, in fact, it was malignant causes much more significant effects for the patient, as opposed to the situation when the benign tumor is diagnosed as cancer.

Table 2.4 Types of classification errors taking into account lack of decision

Actual	Predicted		
	P' (Malignant)	N' (Benign)	NA
P (Malignant)	TP (correct positive decision)	FN (type II error: false negative)	N1 (positive unclassified)
N (Benign)	FP (type I error: false positive)	TN (correct negative decision)	N0 (negative unclassified)

In such models the concept of cost matrix (cost function) is used where for each error type a weight (penalty) is assigned for a wrong decision. The quality value is the sum of costs (penalties) assigned to the classifier for making wrong decisions. Such a cost matrix will be used to evaluate the classification quality of decision modules implemented in the OvaExpert system (see Sect. 4.2).

Chapter 3
Elements of Fuzzy Set Theory

There are many misconceptions about fuzzy logic. The principal misconception is that fuzzy logic is fuzzy. The stated definition underscores that fuzzy logic is precise. In fuzzy logic precision is achieved through association of fuzzy sets with membership functions and, more generally, association of granules with generalized constraints. What this implies is that fuzzy logic is what may be called precisiated logic.

Zadeh [160]

3.1 Fuzzy Sets

The notion of a fuzzy set was first introduced in 1965 by Lotfi A. Zadeh in his work entitled "Fuzzy Sets" (see Zadeh [154]). This work gave birth to a broad and dynamic family of methods and approaches in computer science, mathematics and control theory. Its first applications date back to the 1970s when Ebrahim H. Mamdani (see Mamdani [75]) proposed a method for using fuzzy logic for control. Further dynamic growth of this branch took place in the 1980s when the Japanese started broad-scale applications of fuzzy logic in industry and everyday applications (e.g. controlling metro, cameras, or household appliances). Today, fuzzy controllers are a standard in many household appliances and for example in car industry e.g. ABS or automatic gearbox controllers. Fuzzy logic also finds its wide range applications in decision systems.

The notion of a fuzzy set is a generalization of a crisp set. Let $\mathcal{U}$ be a universe of any elements, finite or not. Set $D \subset U$ can be described uniquely by a characteristic function $1_D : \mathcal{U} \to 0, 1$ such that:

$$1_D(x) = \begin{cases} 1, & \text{if } x \in D, \\ 0, & \text{else.} \end{cases} \tag{3.1}$$

© Springer International Publishing AG 2018   29
K. Dyczkowski, *Intelligent Medical Decision Support System Based on Imperfect Information*, Studies in Computational Intelligence 735,
https://doi.org/10.1007/978-3-319-67005-8_3

The concept of a fuzzy set, on the other hand, lies in gradation of membership of its elements $x \in \mathcal{U}$. This notion is a generalization of the crips set concept where the characteristic function may take any value in the interval $[0, 1]$ instead of $\{0, 1\}$.

Definition 3.1 Fuzzy set in $\mathcal{U}$ is a nebular complex of elements from $\mathcal{U}$, identified with function

$$A : \mathcal{U} \to [0, 1] \tag{3.2}$$

Such a function is called membership function of fuzzy set A.

A fuzzy set may be represented as a set of ordered pairs

$$\{(x, A(x)) : x \in \mathcal{U}\}. \tag{3.3}$$

The number $A(x)$ is interpreted as degree of membership of x to a fuzzy set A. In the same manner as a crisp set $\mathcal{D} \subset \mathcal{U}$ can be identified with its characteristic function $1_{\mathcal{D}}$, a fuzzy set is identified with membership function $A : \mathcal{U} \to [0, 1]$.

3.2 Triangular Operations and Negations

First, we will present elements of triangular norms theory that will be useful in further considerations. Gottward [46, 47], Klement et al. [69] are good examples of a monographs on this issue.

3.2.1 The Notion of a Triangular Norm

The history of triangular norms dates back to the 1940s when Karl Menger pondered over probabilistic metric spaces (see Menger [79]). Triangular norms (t-norms) emerged as a tool for generalizing the triangle inequality into such spaces (thus the name). Later, in the 1960s B. Schweizer and A. Sklar elaborated further on these issues in Schweizer and Sklar [110, 111]. In 1980s researchers noticed that triangular norms are perfectly suitable for numerical interpretation of logical conjunctions in multi-value logics (see Gotwald [46]).

Definition 3.2 A binary operation $t : [0, 1] \times [0, 1] \to [0, 1]$ is called a triangular norm (abbreviated t-norm) if the following conditions are met:

(T1.)	$\forall a, b \in [0, 1]:\ a\,t\,b = b\,t\,a,$	*(commutativity)*
(T2.)	$\forall a, b, c \in [0, 1]:\ (a\,t\,b)\,t\,c = a\,t\,(b\,t\,c),$	*(associativity)*
(T3.)	$\forall a, b, c, d \in [0, 1]:\ (a \leq b\ \&\ c \leq d) \Rightarrow a\,t\,c \leq b\,t\,d,$	*(monotonicity)*
(T4.)	$\forall a \in [0, 1]:\ a\,t\,1 = a.$	$(1 - $ *neutral element*$)$

Definition 3.3 A binary operation $s : [0, 1] \times [0, 1] \to [0, 1]$ is called a triangular conorm (abbreviated t-conorm) if s meets **T1.–T3.** and the following condition

$$(\textbf{T5.}) \ \forall u \in [0, 1] : \ a \, s \, 0 = a. \ (0 - \textit{neutral element})$$

t-norms and t-conorms shall be jointly referred to as *triangular operations (t-operations)*. Basic examples of t-operations include:

1. minimum t-norm $\wedge$
$$a \wedge b := min(a, b),$$

2. maximum t-conorm $\vee$
$$a \vee b := max(a, b),$$

3. drastic t-norm t_d
$$a \, t_d \, b := \begin{cases} a \wedge b, & \text{if } a \vee b = 1, \\ 0 & \text{else}, \end{cases}$$

4. drastic t-conorm s_d
$$a \, s_d \, b := \begin{cases} a \vee b, & \text{if } a \wedge b = 0, \\ 1 & \text{else}, \end{cases}$$

5. algebraic t-norm t_a
$$a \, t_a \, b := ab,$$

6. algebraic t-conorm s_a
$$a \, s_a \, b := a + b - ab,$$

7. Łukasiewicz t-norm t_L
$$a \, t_L \, b := 0 \vee (a + b - 1),$$

8. Łukasiewicz t-conorm s_L
$$a \, s_L \, b := 1 \wedge (a + b).$$

For each t-norm t and t-conorm s and $a, b, c \in [0, 1]$ the following properties are met:

(a) $a \, t \, 0 = 0, a \, s \, 1 = 1$,
(b) $a \, t_d b \leq a \, t \, b \leq a \wedge b \leq a \vee b \leq a \, s \, b \leq a \, s_d b$,
(c) $a \, t \, a \leq a \leq a \, s \, a$,
(d) $a \, t \, b = 1 \Leftrightarrow a = b = 1$,
(e) $a \, s \, b = 0 \Leftrightarrow a = b = 0$,

(f) $(\forall a \in [0, 1]: a\,t\,a = a) \Leftrightarrow t = \wedge,$
(g) $(\forall a \in [0, 1]: a\,s\,a = a) \Leftrightarrow s = \vee,$
(h) $(\forall a, b, c \in [0, 1]: a\,t\,(b\,s\,c) = (a\,t\,b)\,s\,(a\,t\,c)) \Leftrightarrow s = \vee,$
(i) $(\forall a, b, c \in [0, 1]: a\,s\,(b\,t\,c) = (a\,s\,b)\,t\,(a\,s\,c)) \Leftrightarrow t = \wedge.$

The property (b) can be written shorter as

$$t_d \leq t \leq \wedge \leq \vee \leq s \leq s_d, \tag{3.4}$$

where the partial order relation $\leq$ is defined in the following manner:

$$\boldsymbol{u} \leq \boldsymbol{v} \Leftrightarrow \forall a, b \in [0, 1]: a\,\boldsymbol{u}\,b \leq a\,\boldsymbol{v}\,b. \tag{3.5}$$

Thus, t_d i $\wedge$ are extreme t-norms, and s_d i $\vee$ are extreme t-conorms. Properties (f) and (g) tell us that $\wedge$ is the only idempotent t-norm and $\vee$ is the only idempotent t-conorm.

There is bijective correspondence between t-norms and t-conorms. More specifically:

(a) If a binary operation t is a t-norm, then operation t^* such that

$$\forall a, b \in [0, 1]: a\,t^*\,b := 1 - (1 - a)\,t\,(1 - b) \tag{3.6}$$

is a t-conorm.

(b) If a binary operation s is a t-conorm, then operation s^* such that

$$\forall a, b \in [0, 1]: a\,s^*\,b := 1 - (1 - a)\,s\,(1 - b) \tag{3.7}$$

is a t-norm.

(c)

$$(t^*)^* = t, \quad (s^*)^* = s. \tag{3.8}$$

Please note that $t = s^*$ implies $t^* = (s^*)^* = s$ and the other way round. If $t = s^*$ or (identically) $s = t^*$, we say that t and s are *dual*. Basic examples of dual t-operations include: $\wedge$ and $\vee$, t_a and s_a, t_L and s_L, t_d and s_d.

A continuous t-norm t is Archimedean if and only if

$$a\,t\,a < a \tag{3.9}$$

for all $a \in (0, 1)$.

We say that a continuous t-norm is *strict*, if it is strictly increasing in $(0, 1) \times (0, 1)$ i.e. if $a < b \Rightarrow a\,t\,c < b\,t\,c$. Strictness t thus means that it is continuous and strictly monotonic with respect to both arguments. Each strict t-norm is Archimedean as strictness t implies that $a\,t\,a < a\,t\,1 = a$.

3.2.2 Basic Families of Triangular Operations

1. Schweizer t-norms with $\lambda > 0$:

$$a\, t_{S,\lambda}\, b := (0 \vee (a^\lambda + b^\lambda - 1))^{\frac{1}{\lambda}},$$

$$a\, s_{S,\lambda}\, b := 1 - (0 \vee ((1-a)^\lambda + (1-b)^\lambda - 1))^{\frac{1}{\lambda}};$$

2. Yager t-norms with $\lambda \geq 1$:

$$a\, t_{Y,\lambda}\, b := 1 - (1 \wedge ((1-a)^\lambda + (1-b)^\lambda)^{\frac{1}{\lambda}}),$$

$$a\, s_{Y,\lambda}\, b := 1 \wedge (a^\lambda + b^\lambda)^{\frac{1}{\lambda}};$$

3. Hamacher t-norms with $\lambda \geq 0$:

$$a\, t_{H,\lambda}\, b := \frac{ab}{\lambda + (1-\lambda)(a + b - ab)},$$

$$a\, s_{H,\lambda}\, b := \frac{a + b - ab - (1-\lambda)ab}{1 - (1-\lambda)ab};$$

4. Frank t-norms with $\lambda > 0, \lambda \neq 1$:

$$a\, t_{F,\lambda}\, b := \log_\lambda \left(1 + \frac{(\lambda^a - 1)(\lambda^b - 1)}{\lambda - 1}\right),$$

$$a\, s_{F,\lambda}\, b := 1 - \log_\lambda \left(1 + \frac{(\lambda^{1-a} - 1)(\lambda^{1-b} - 1)}{\lambda - 1}\right);$$

5. Weber t-norms, for $\lambda > -1$

$$a\, t_{W,\lambda}\, b := 0 \vee \left(\frac{a + b - 1 + \lambda ab}{1 + \lambda}\right),$$

$$a\, s_{W,\lambda}\, b := 1 \wedge \left(\frac{(1+\lambda)(a+b) - \lambda ab}{1 + \lambda}\right).$$

Please note that, for example

$$t_a = t_{H,1}, \quad s_a = s_{H,1},$$

$$t_L = t_{S,1}, \quad s_L = s_{S,1}.$$

Also properties of Frank t-norms are worth stressing:

$$a \ t_{F,\lambda}b \xrightarrow[\lambda \to 0]{} a \wedge b, \quad a \ s_{F,\lambda}b \xrightarrow[\lambda \to 0]{} a \vee b,$$

$$a \ t_{F,\lambda}b \xrightarrow[\lambda \to 1]{} a \ t_a b, \quad a \ s_{F,\lambda}b \xrightarrow[\lambda \to 1]{} a \ s_a b,$$

$$a \ t_{F,\lambda}b \xrightarrow[\lambda \to \infty]{} a \ t_{\text{Ł}}b, \quad a \ s_{F,\lambda}b \xrightarrow[\lambda \to \infty]{} a \ s_{\text{Ł}}b.$$

3.2.3   Negations

When considering fuzzy set theory, the notion of negation plays a vital role.

Definition 3.4 Function $\nu : [0, 1] \to [0, 1]$ is called negation whenever ν is nonincreasing and meets conditions $\nu(0) = 1$ and $\nu(1) = 0$.

Definition 3.5 Negation ν is called strict negation if it is strictly decreasing and continuous.

Definition 3.6 Negation ν is called strong negation if it is strict and involutive i.e.

$$\nu(\nu(a)) = a \ \text{ for each } \ a \in [0, 1]. \tag{3.10}$$

Note that an inverse function ν^{-1} to any strict negation ν is also a strict negation. Strong negation is identical to its inverse.

The minimum negation is negation ν_* such that

$$\nu_*(a) := \begin{cases} 1 & \text{if } a = 0, \\ 0 & \text{otherwise} \end{cases} \tag{3.11}$$

and the maximum negation is ν^* where

$$\nu^*(a) := \begin{cases} 1 & \text{if } a < 1, \\ 0 & \text{otherwise.} \end{cases} \tag{3.12}$$

Therefore, $\nu_* \leq \nu \leq \nu^*$ for any negation ν, where relation $\leq$ is defined as $\nu \leq \mu \Leftrightarrow \forall a \in [0, 1] : \nu(a) \leq \mu(a)$.

Łukasiewicz negation $\nu_{\text{Ł}}$ constitutes an important example of strong (thus also strict) negation with

$$\nu_{\text{Ł}}(a) := 1 - a \quad \text{for each } \ a \in [0, 1]. \tag{3.13}$$

Another example of a strong negation includes functions introduced by Sugeno:

$$\nu_{S,\lambda}(a) := \frac{1 - a}{1 + \lambda a} \quad \text{for parameter } \ \lambda > -1. \tag{3.14}$$

Further, we will write *Sneg* for the whole family of strong negations.
Please note that each strong negation $\nu \in Sneg$ has exacly one fixed point $a^* \in (0, 1)$.
For example $a^* = 0.5$ for $\nu = \nu_L$. In general, $\nu(a^*) = a^*$ and

$$\forall a \neq a^* : \ \nu(a) < a^* < a \ \text{ or } \ a < a^* < \nu(a), \tag{3.15}$$

i.e.

$$\forall a \in [0, 1] : \ a \vee \nu(a) \geq a^*. \tag{3.16}$$

3.3 Basic Operations on Fuzzy Sets

Triangular operations and negations find their application in defining operations on
fuzzy sets (see Gottwald [47], Klement et al. [69]). Let t denote any t-norm, s –
t-conorm, ν – negation and $A, B \in [0, 1]^{\mathcal{U}}$. Then, *intersection* $A \cap_t B$ of fuzzy sets A
and B induced by t, *union* $A \cup_s B$ induced by s and *Cartesian product* $A \times_t B$ induced
by t are defined as follows:

$$(A \cap_t B)(x) := A(x) \ t \ B(x) \quad \text{for each } x \in \mathcal{U},$$

$$(A \cup_s B)(x) := A(x) \ s \ B(x) \quad \text{for each } x \in \mathcal{U}, \tag{3.17}$$

$$(A \times_t B)(x, y) := A(x) \ t \ B(y) \quad \text{for each } (x, y) \in \mathcal{U}_1 \times \mathcal{U}_2.$$

In the last case we assume $A \in [0, 1]^{\mathcal{U}_1}$ and $B \in [0, 1]^{\mathcal{U}_2}$.
Complement A^ν fuzzy set A induced by ν is defined as

$$A^\nu(x) := \nu(A(x)) \quad \text{for each } x \in \mathcal{U}. \tag{3.18}$$

If $t = \wedge, s = \vee$ i $\nu = \nu_L$, we will use simplified notation and terminology:

$$A \cap B := A \cap_\wedge B, \quad (intersection)$$

$$A \cup B := A \cup_\vee B, \quad (union) \tag{3.19}$$

$$A \times B := A \times_\wedge B, \quad (cartesian \ product)$$

$$A' := A^{\nu_L}. \quad (complement \ A)$$

Operations $\cap, \cup, \times$ and $'$ are standard operations on fuzzy sets introduced by Zadeh
[154].
  Inclusion $\subset$ and equality $=$ relations for two fuzzy sets are also defined in a
standard manner:

$$A \subset B \ \Leftrightarrow \ \forall x \in \mathcal{U} : \ A(x) \leq B(x), \tag{3.20}$$

$$A = B \iff A \subset B \text{ and } B \subset A. \tag{3.21}$$

Directly from Definitions 3.2, 3.3 and (3.17) we conclude that operations $\cap_t$ i $\cup_s$ are commutative, associative, monotonic and $1_{\mathcal{U}}$, $1_\emptyset$ are their respective identity elements. The following properties apply for $A, B, C \in [0, 1]^{\mathcal{U}}$:

$$A \cap_{t_d} B \subset A \cap_t B \subset A \cap B \subset A, B \subset A \cup B \subset A \cup_s B \subset A \cup_{s_d} B,$$

$$A \cap_t A \subset A \subset A \cup_s A,$$

$$A \cap A = A \cup A = A, \tag{3.22}$$

$$A \cap_t (B \cup C) = (A \cap_t B) \cup (A \cap_t C),$$

$$A \cup_s (B \cap C) = (A \cup_s B) \cap (A \cup_s C).$$

Moreover, de Morgan laws apply:

$$(A \cap_t B)' = A' \cup_{t^*} B',$$

$$(A \cup_s B)' = A' \cap_{s^*} B'.$$

However, $A \cap A' \neq 1_\emptyset$ and $A \cup A' \neq 1_{\mathcal{U}}$ for $A \notin \{0, 1\}^{\mathcal{U}}$.
If ν is strong, then
$$(A^\nu)^\nu = A.$$

Set
$$A_t := \{x \in \mathcal{U} : A(x) \geq t\}, \quad t \in (0, 1],$$

is called *t-cut* of fuzzy set A, and

$$A^t := \{x \in \mathcal{U} : A(x) > t\}, \quad t \in [0, 1),$$

is called *sharp t-cut* A. Therefore, $A_t = A^{-1}([t, 1])$ and $A^t = A^{-1}((t, 1])$. It is easy to note that
$$A \subset B \implies \forall t : A_t \subset B_t,$$

$$t \leq u \implies A_u \subset A_t,$$

$$(A * B)_t = A_t * B_t,$$

where $*$ is any of the operations $\cap, \cup, \times$.
Identical properties apply to sharp *t*-cuts. Moreover,

$$A = B \iff \forall t \in (0, 1] : A_t = B_t \iff \forall t \in (0, 1] : A^t = B^t$$

Let us define

$$core(A) := A_1, \quad (core\ of\ fuzzy\ set\ A)$$

and

$$supp(A) := A^0. \quad (support\ of\ fuzzy\ set\ A)$$

If $supp(A)$ is finite, we call A a *finite* fuzzy set. The family of all such sets in $\mathcal{U}$ will be denoted by *FFS*. *FCS* denotes the family *of all finite sets* in $\mathcal{U}$.

Fuzzy sets with a one-element support are called *singletons*. The symbol a/x denotes a singleton with $a > 0$ in point $x \in \mathcal{U}$. Each fuzzy set can be represented as a sum of singletons:

$$A = \bigcup_{x \in supp(A)} A(x)/x. \tag{3.23}$$

If A is finite and $supp(A) = \{x_1, x_2, \ldots, x_n\}$, $n \geq 1$, we will use the following notation:

$$A = A(x_1)/x_1 + A(x_2)/x_2 + \ldots + A(x_n)/x_n.$$

We say that $A, B \in [0, 1]^{\mathcal{U}}$ are *disjoint* if $A \cap B = 1_\emptyset$. If $\mathcal{U}$ is linearly ordered by a given relation $\leq$, then A is called a *convex* fuzzy set if

$$\forall x, y, z \in \mathcal{U}\ (x \leq y \leq z) : A(y) \geq A(x) \wedge A(z).$$

3.4 Cardinality of a Fuzzy Set

Similarly to a regular set, cardinality of a fuzzy set is one of its basic characteristics. In the case of fuzzy sets this notion is strongly motivated by applications. Questions like "How many x's are p", "Are there more x's that are p than x's that are q", where p, q are fuzzy characteristics, are all questions about cardinalities of fuzzy sets or comparisons of these cardinalities. Getting adequate answers to such questions is crucial for example in decision making with fuzzy information, communication with data bases at natural conceptual level, modelling natural language expressions etc. (see e.g. Dubois and Prade [33], Delgado et al. [28], Lin and Kerre [74], Ralescu [109]).

It is clear that when defining cardinality of a fuzzy set A the main difficulty and difference in comparison to a crisp set is that membership of element $x \in \mathcal{U}$ in A is gradable.

We will now move to presenting contemporary approaches to defining cardinality $|A|$ of a fuzzy set A. An extensive monographic overview of approaches to cardinality of fuzzy sets and their extensions can be found in monographs Wygralak [149, 150]. Because from application point of view, these are finite fuzzy sets that play a dominant role, in this overview we will basically limit ourselves to such fuzzy sets. Let us thus assume that $A \in FFS$, $n := |supp(A)|$ and $m := |core(A)|$.

3.4.1 Scalar Approach

The first, operationally and computationally simple approach is the scalar approach in which cardinality of a fuzzy set A is understood as a single real, nonnegative number. The most frequently used scalar cardinalities include:

(i) $|A| := sc(A) = \displaystyle\sum_{x \in supp(A)} A(x),$ (i.e. *sigma count* of a fuzzy set A),

(see De Luca and Termini [26, 27], Zadeh [156–159])

(ii) $|A| := sc_p(A) = \displaystyle\sum_{x \in supp(A)} (A(x))^p, \quad p > 0,$

(see Dubois and Prade [33], Kaufmann [68])

(iii) $|A| := |A_t|$ or $|A| := |A^t|,$ (Ralescu [109], Wygralak [147])

and in particular

$|A| := |core(A)|$ or $|A| := |supp(A)|.$ (Gotwald [48])

In [147] a general, axiomatic approach to scalar cardinalities has been formulated.

Definition 3.7 Function $\sigma : FFS \to [0, \infty)$ is called scalar cardinality , if σ meets the following conditions for any $a, b \in [0, 1]$, $A, B \in FFS$ and $x, y \in \mathcal{U}$:
(SC1) $\sigma(1/x) = 1$,
(SC2) $a \leq b \Rightarrow \sigma(a/x) \leq \sigma(b/y)$,
(SC3) $A \cap B = 1_\emptyset \Rightarrow \sigma(A \cup B) = \sigma(A) + \sigma(B)$.
If σ meets the three postulates above, then $\sigma(A)$ is called scalar cardinality of a fuzzy set A.

Definition 3.8 Function $f : [0, 1] \to [0, 1]$ is called cardinality pattern if it meets the following conditions:

1. is nondecreasing i.e. $\forall_{a,b \in [0,1]} f(a) \leq f(b)$ if $a \leq b$,
2. and meets limit conditions $f(0) = 0$ i $f(1) = 1$.

Using the notion of cardinality pattern we can define *sigma f-Count* cardinality.

Definition 3.9 Function $sc_f : FFS \to [0, \infty)$ is called cardinality of the *sigma f-Count* type and is defined as

$$\forall A \in FFS : \quad sc_f(A) = \sum_{x \in supp(A)} f(A(x)), \tag{3.24}$$

where f is a function meeting cardinality pattern conditions from Definition 3.8.

By Definition 3.8, weighting function f (*cardinality pattern*) expresses our understanding of a scalar cardinality of singleton i.e. it interprets how important a given element is for cardinality of a given fuzzy set. By appropriate adjustments of this pattern, we may generate the above-mentioned classical scalar cardinalities i.e.

1. $sc_f(A) = sc(A)$, when $f = id$,
2. $sc_f(A) = sc_p(A)$, when $f(a) = a^p$ for $a \in [0, 1]$,
3. $sc_f(A) = |A_t|$, when $f(a) = \begin{cases} 1 & \text{for } a \geq t, \\ 0 & \text{otherwise.} \end{cases}$

For the purposes of this book let us define five basic families of cardinality patterns. Each of them interprets membership of specific elements in a fuzzy set in a different manner and, what follows, it may generate different cardinalities of this set. For the sake of completeness, let us stress that identity function (*id*) meets the cardinality pattern conditions.

1. $f_{1,t,p}$, where $t \in (0, 1]$ and $p \geq 0$. As an effect, use of a function of this type yields cardinality called *counting by thresholding* Wygralak [150] (see Fig. 3.1)

$$f_{1,t}(x) = \begin{cases} 1, & \text{if } x \geq t, \\ x^p & \text{otherwise.} \end{cases} \tag{3.25}$$

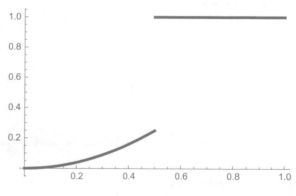

Fig. 3.1 Cardinality pattern $f_{1,t,p}$ for $t = 0.5$ and $p = 2$

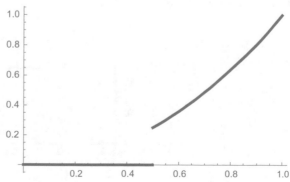

Fig. 3.2 Cardinality pattern $f_{2,t,p}$ for $t = 0.5$ and $p = 2$

2. $f_{2,t,p}$, where $t \in (0, 1]$ and $p \geq 0$. As an effect, use of a function of this type yields cardinality called *counting by thresholding and joining* (see Wygralak [150]). An example of such a function has been presented in the Figs. 3.2)

$$f_{2,t}(x) = \begin{cases} x^p, & \text{if } x \geq t, \\ 0 & \text{otherwise.} \end{cases} \tag{3.26}$$

3. $f_{3,t_1,t_2,p}$, where $t_1, t_2 \in [0, 1]$, $t_1 \leq t_2$ and $p \geq 0$. (see Fig. 3.3)

$$f_{3,t_1,t_2,p} = \begin{cases} 1, & \text{if } x \geq t_2, \\ x^p, & \text{if } x \in (t_1, t_2), \\ 0 & \text{otherwise.} \end{cases} \tag{3.27}$$

4. f_{4,t_1,t_2}, where $t_1, t_2 \in [0, 1]$ and $t_1 \leq t_2$. (see Fig. 3.4)

$$f_{4,t_1,t_2}(x) = \begin{cases} 1, & \text{if } x \geq t_2, \\ 0.5, & \text{if } x \in (t_1, t_2) \\ 0 & \text{otherwise} \end{cases} \tag{3.28}$$

Fig. 3.3 Cardinality pattern $f_{3,t_1,t_2,p}$ for $t_1 = 0.3$, $t_2 = 0.7$ and $p = 1$

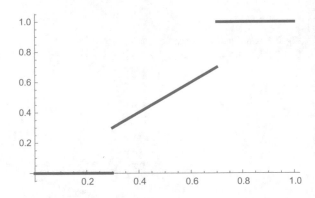

Fig. 3.4 Cardinality pattern f_{4,t_1,t_2} for $t_1 = 0.3$, $t_2 = 0.7$

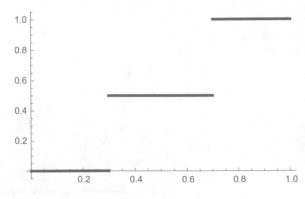

5. $f_{5,t,p}$, where $t > 0$ and $p \in (0, 1)$. (see Fig. 3.5)

$$f_{5,t,p}(x) = \frac{1}{1 + e^{(-t(x-p))}} \qquad (3.29)$$

Example 3.1 For a fuzzy set:

$$A = 0.59/x_1 + 0.63/x_2 + 0.38/x_3 + 0.36/x_4 + 0.33/x_5 + 0.09/x_6. \qquad (3.30)$$

we receive the following values of *sigma f-Count* for various cardinality patterns:

- $sc_f(A) = 2.38$ for identity function $f = id$—classical sigma count non-weighting for membership function values,
- $sc_f(A) = 2$ for a cardinality pattern (3.25)—this function promotes high values (exceeding threshold) by changing them to one,
- $sc_f(A) = 0.74$ for a cardinality pattern (3.26)—this function is very restrictive as it rejects (reduces to 0) all the values of membership function below threshold, at the same time if the parameter p is greater than unity, it decreases all the values of membership function (except 1).
- $sc_f(A) = 2.29$ for a cardinality pattern (3.27)—the function rejects the values of membership functions below the first threshold and increases the values exceeding the second threshold to one,
- $sc_f(A) = 2.5$ for a cardinality pattern (3.28)—this function operates similarly to (3.27) with such a difference that it significantly levels out the values from the interval (t_1, t_2) by dividing the values of membership function into three classes with various influence on the value of cardinality,
- $sc_f(A) = 2.01$ for a cardinality pattern (3.29)—this function does not reject any values of the membership function, but it only highlights them. Its operation is similar to contrast enhancement function.

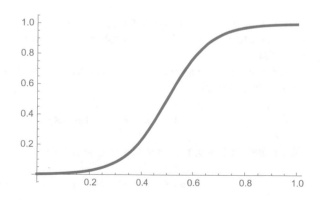

Fig. 3.5 Cardinality pattern $f_{5,t,p}$ for $t = 0.5, p = 12$

As can be seen, weighting functions (cardinality patterns) play a essential role in calculating cardinalities of fuzzy sets and their correct selection for a given problem is of vital importance.

3.4.2   Fuzzy Approach

In this approach cardinality of a fuzzy set is understood as a convex fuzzy set in $\mathbb{N}$. In this case it is thus a convex set of fuzzy cardinalities in a usual sense.

At the expense of its complexity, the fuzzy approach offers a far more adequate description of a fuzzy set cardinality. Convex fuzzy sets of cardinal numbers (in particular: natural numbers) expressing cardinalities of fuzzy sets in $\mathcal{U}$ are called *generalized cardinal numbers*. We will denote them with lowercase letters $\alpha, \beta, \gamma, \ldots$ from the beginning of the Greek alphabet. The equality $\alpha = \beta$ should be understood as $\alpha(i) = \beta(i)$ for each i. Before we move to presentation of basic types of generalized cardinal numbers of finite fuzzy sets, we will introduce some additional symbols and notation conventions.

Let $A \in FFS$ and

$$[A]_i := \bigvee \{t \in (0, 1] : \ |A_t| \geq i\} \ \text{ for } \ i \in \mathbb{N}. \tag{3.31}$$

Please note that the symbol $[A]_i$ also makes sense for infinite A and transfinite cardinal number i.

We can see that

$$[A]_0 = 1, \ \ [A]_i = 0 \ \text{ for } \ i > n, \ \ [A]_i = 1 \ \text{ for } \ i \leq m$$

and

$$A \subset B \ \Rightarrow \ \forall i : \ [A]_i \leq [B]_i.$$

If $0 < i \leq n$, we have $[A]_i \in (0, 1]$ and $[A]_i$ is the i-th element in a nonincreasing sequence of all positive membership degrees $A(x)$, taking into account possible repetitions.

We will write generalized cardinal numbers of finite fuzzy sets with vectors and the following notation:

$$\alpha = (a_0, a_1, \ldots, a_k, (a)) \ \text{ for } \ k \in \mathbb{N}$$

means that $\alpha(i) = a_i$ for $i \leq k$ and $\alpha(i) = a$ for $i > k$. Moreover,

$$(a_0, a_1, \ldots, a_k) := (a_0, a_1, \ldots, a_k, (0)).$$

For example, notation $\alpha = (0, 0.3, 0.4, 0.7, 0.2)$ means that $\alpha(0) = 0$, $\alpha(1) = 0.3$, $\alpha(2) = 0.4$, $\alpha(3) = 0.7$, $\alpha(4) = 0.2$ and $\alpha(i) = 0$ for $i > 4$.

Let us now present three classes of generalized cardinal numbers that are the most important ones from both theoretical and application point of view.

Definition 3.10 Let $A \in FFS$ and $\alpha = |A|$.

1. Generalized cardinal number *FGCount* is defined as

$$\alpha(k) := [A]_k \quad \text{for } k \in \mathbb{N}, \tag{3.32}$$

i.e.

$$\alpha = (1, [A]_1, [A]_2, \ldots, [A]_n). \tag{3.33}$$

Number $\alpha(k)$ is interpreted as a degree to which A has at least k elements, thus α in fact constitutes lower estimate of the cardinality A.

2. Generalized cardinal number *FLCount* is defined as

$$\alpha(k) := 1 - [A]_{k+1} \quad \text{for } k \in \mathbb{N}, \tag{3.34}$$

i.e.

$$\alpha = (1 - [A]_1, 1 - [A]_2, \ldots, 1 - [A]_n, (1)). \tag{3.35}$$

In this case we interpret $\alpha(k)$ as a degree to which A includes at most k elements; α is thus the upper estimate of cardinality A.

3. Generalized cardinal number *FECount* is defined as

$$\alpha(k) := [A]_k \wedge (1 - [A]_{k+1}) \quad \text{for } k \in \mathbb{N}, \tag{3.36}$$

i.e.

$$\alpha = (1 - [A]_1, 1 - [A]_2, \ldots, 1 - [A]_l, [A]_l, [A]_{l+1}, \ldots, [A]_n). \tag{3.37}$$

with $l = \min\{k : [A]_k + [A]_{k+1} \le 1\}$. $\alpha(k)$ expresses the degree to which A has exactly k elements. α is the intersection of generalized cardinal numbers (3.32) and (3.34). It may be perceived as the "actual" generalized cardinal number of a fuzzy set A.

Generalized cardinal numbers of the type *FGCount* and *FLCount* have been introduced by Zadeh in Zadeh [157, 158], and *FECount* by Zadeh in Zadeh [158] and Wygralak in Wygralak [146].

Further in this book, we will denote generalized cardinal number α of type *FGCount* of fuzzy set A as $FG(A)$, number of type FLCount we will denote as $FL(A)$ and type *FECount* it will be $FE(A)$.

In some cases (especially in the case of *FECount* and its use in decision making algorithms) there is a necessity of defuzzification of such a cardinal number (e.g.

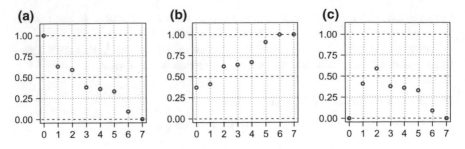

Fig. 3.6 Generalized cardinal numbers of set A (3.39): **a** $FG(A)$, **b** $FL(A)$, **c** $FE(A)$

at the final decision stage). The most common way is to calculate centre of gravity (COG) of a figure defined by axis x and values of the cardinal number. In the case of generalized cardinal number FE for a finite fuzzy set in universe $\mathcal{U}$ we can calculate it with the following formula:

$$COG(\alpha) = \frac{\sum_{l=1}^{|\mathcal{U}|} l * \alpha(l)}{\sum_{l=1}^{|\mathcal{U}|} \alpha(l)}. \tag{3.38}$$

We will use the example below to present individual types of fuzzy set cardinalities.

Example 3.2 For a fuzzy set

$$A = 0.59/x_1 + 0.63/x_2 + 0.38/x_3 + 0.36/x_4 + 0.33/x_5 + 0.09/x_6 \tag{3.39}$$

generalized cardinal numbers are the following (see Fig. 3.6) (with COG value calculated for FE(A)):

- $FG(A) = (1, 0.63, 0.59, 0.38, 0.36, 0.33, 0.09, 0)$,
- $FL(A) = (0.37, 0.41, 0.62, 0.64, 0.67, 0.91, 1.00, 1.00)$,
- $FE(A) = (0.00, 0.41, 0.59, 0.38, 0.36, 0.33, 0.09, 0.00)$;
  $COG(FE(A)) = 2.94$.

As noted by Wygralak [148], definitions of generalized cardinal numbers from Definition 3.10 are only appropriate for fuzzy sets with standard operations $\cap$ and $\cup$. Now, based on Wygralak [148] we will present their generalizations to fuzzy sets with operations induced by t-norms. Let t designate any t-norm, and let ν designate negation.

Definition 3.11   1. *Generalized FGCount of a fuzzy set A* is defined by formula:

$$\alpha(k) := [A]_1 \, t \, [A]_2 \, t \, \ldots \, t \, [A]_k \quad \text{for } k \in \mathbb{N}. \tag{3.40}$$

Using vector notation, we can write generalized *FGCount* of a fuzzy set A as:

$$\alpha = (1, [A]_1, [A]_1 \ t \ [A]_2, \ldots, [A]_1 \ t \ [A]_2 \ t \ \ldots \ t \ [A]_n). \qquad (3.41)$$

Please note that $\alpha(k)$ is a degree to which A has (at least) i elements for each $i \leq k$. By substituting $t = \wedge$, we obtain classical *FGCount* of fuzzy set A.

2. *Generalized FLCount* of a fuzzy set A is defined by formula:

$$\alpha(k) := \nu([A]_{k+1}) \ t \ \nu([A]_{k+2}) \ t \ \ldots \ t \ \nu([A]_n) \ \text{dla} \ k \in \mathbb{N}. \qquad (3.42)$$

In this case $\alpha(k)$ is the degree to which A includes at most i elements for each $i \geq k$. In vector notation we have:

$$\begin{aligned} \alpha = (&\nu([A]_1) \ t \ \nu([A]_2) \ t \ \ldots \ t \ \nu([A]_n), \qquad (3.43) \\ &\nu([A]_2) \ t \ \nu([A]_3) \ t \ \ldots \ t \ \nu([A]_n), \\ &\qquad\qquad \vdots \\ &\nu([A]_{n-1} \ t \ \nu([A]_n), \\ &\nu([A]_n), (1)). \end{aligned}$$

By substituting $t = \wedge$ i $\nu = \nu_L$ we obtain usual *FLCount*.

3. *Generalized FECount of a fuzzy set A* is defined by formula:

$$\begin{aligned} \alpha(k) := \ &[A]_1 \ t \ [A]_2 \ t \ \ldots \ t \ [A]_k \ t \qquad (3.44) \\ &\nu([A]_{k+1}) \ t \ \nu([A]_{k+2}) \ t \ \ldots \ t \ \nu([A]_n) \ \text{ for } \ k \in \mathbb{N}, \end{aligned}$$

and in vector notation:

$$\begin{aligned} \alpha = (&\nu([A]_1) \ t \ \nu([A]_2) \ t \ \ldots \ t \ \nu([A]_n), \qquad (3.45) \\ &[A]_1 \ t \ \nu([A]_2) \ t \ \nu([A]_3) \ t \ \ldots \ t \ \nu([A]_n), \\ &[A]_1 \ t \ [A]_2 \ t \ \nu([A]_3) \ t \ \ldots \ t \ \nu([A]_n), \\ &\qquad\qquad \vdots \\ &[A]_1 \ t \ [A]_2 \ t \ \ldots \ t \ [A]_{n-1} \ t \ \nu([A]_n), \\ &[A]_1 \ t \ [A]_2 \ t \ \ldots \ t \ [A]_n). \end{aligned}$$

Generalized *FECount* α of a fuzzy set A induced by t and ν, as defined in (3.44), is an intersection of generalized cardinal numbers (3.40) and (3.42).

If $t = \wedge$ and $\nu = \nu_L$, α can be reduced to a generalized cardinal number of type *FECount* defined in (3.36). With $t = \wedge$ and $\nu = \nu^*$ the formula (3.45) takes the form of

$$\alpha(k) = (\underbrace{0, 0, \ldots, 0}_{m}, 1, [A]_{m+1}, [A]_{m+2}, \ldots, [A]_n), \qquad (3.46)$$

thus α becomes generalized cardinal number of Dubois type as introduced by [33].

Further, by $GFE_{t,\nu}$ we will designate a family of generalized cardinal numbers (3.45) from $A \in FFS$, induced by t i ν. In order to stress that $\alpha \in GFE_{t,\nu}$ if needed we will also use indexed notation $\alpha^{t,\nu}$ in the place of α.

Theorem 3.1 *Let $A \in FFS$ i $|A| = \alpha \in GFE_{t,\nu}$, where t designates any t-norm, and ν denotes negation. Then*

(a) α is a convex fuzzy set,
(b) if $\nu(a) = 1 \Leftrightarrow a = 0$, then α is normal if and only if $A \in FCS$,
(c) $\alpha = 1_{\{n\}}$, if $A \in FCS$,
(d) $\alpha(k) = 0$ for $k \notin [m, n]$,
(e) if $t \leq \mu$ i $\nu \leq \xi$, then $\alpha^{t,\nu} \subset \alpha^{\mu,\xi}$.

Proof of this theorem has been provided in Wygralak [148].

Addition and multiplication of generalized cardinal numbers in the form (3.41), (3.43) and (3.45) with Archimedean t-norm (including $t = \wedge$), and strict negation ν are defined in a classical manner:

(i) $\alpha + \beta := |A \cup B|$,
    where $A, B \in FFS$ are any disjoint fuzzy sets such that $|A| = \alpha$ and $|B| = \beta$.
(ii) $\alpha \cdot \beta := |A \times B|$,
    where $A, B \in FFS$ are such that $|A| = \alpha$ and $|B| = \beta$.

Addition can also be performed with the use of extension rule:

$$(\alpha + \beta)(k) = \bigvee_{i+j=k} \alpha(i) \, t \, \beta(j) \quad \text{for } k \in \mathbb{N}.$$

Detailed analysis of generalized cardinal numbers (3.41), (3.43) and (3.45) and their arithmetic have been presented in Wygralak [148].

3.4.3 Fuzzy Approach with the Use of Cardinality Patterns

Similarly as in the case of scalar cardinalities, introduction of weighting functions proved useful with fuzzy cardinalities. In many real-life decision problems it is required for modelling to determine the influence of specific elements on cardinality of a fuzzy set.

By using *cardinality patterns* defined in Definition 3.8 we can define *fuzzy f-weighted cardinalities*. Similarly as in the case of previous definitions, let t be any t-norm, let ν be any strong negation, and let f be any cardinality pattern meeting the conditions of Definition 3.8.

Definition 3.12 1. *Generalized f-weighted FGCount (f-FGCount) of a fuzzy set A is defined as:*

$$FG_f(k) := f([A]_1) \, t \, f([A]_2) \, t \, \ldots \, t \, f([A]_k) \quad \text{for } k \in \mathbb{N}. \tag{3.47}$$

Using vector notation, we can write *generalized f-weighted FGCount* of a fuzzy set A as:

$$FG_f(A) = (1, f([A]_1), f([A]_1) \ t \ f([A]_2), \ldots \tag{3.48}$$
$$\ldots, f([A]_1) \ t \ f([A]_2) \ t \ \ldots \ t \ f([A]_n)).$$

By substituting $t = \wedge$ and $f = id$, we obtain classical *FGCount* of fuzzy set A.

2. *Generalized f-weighted FLCount (f-FLCount)* of a fuzzy set A is defined as:

$$FL_f(k) := \nu(f([A]_{k+1})) \ t \ \nu(f([A]_{k+2})) \ t \ \ldots \tag{3.49}$$
$$\ldots \ t \ \nu(f([A]_n)) \ \text{dla} \ k \in \mathbb{N}.$$

In vector notation:

$$FL_f(A) = (\nu(f([A]_1)) \ t \ \nu(f([A]_2)) \ t \ \ldots \ t \ \nu(f([A]_n)), \tag{3.50}$$
$$\nu(f([A]_2)) \ t \ \nu(f([A]_3)) \ t \ \ldots \ t \ \nu(f([A]_n)),$$
$$\vdots$$
$$\nu(f([A]_{n-1}) \ t \ \nu(f([A]_n)),$$
$$\nu(f([A]_n)), (1)).$$

By substituting $t = \wedge$, $\nu = \nu_{\text{Ł}}$ and $f = id$ we obtain *FLCount*.

3. *Generalized f-weighted FECount (f-FECount)* of a fuzzy set A is defined as:

$$FE_f(k) := f([A]_1) \ t \ f([A]_2) \ t \ \ldots \ t \ f([A]_k) \ t \tag{3.51}$$
$$\nu(f([A]_{k+1})) \ t \ \nu(f([A]_{k+2})) \ t \ \ldots \ t \ \nu(f([A]_n)) \ \text{for} \ k \in \mathbb{N},$$

in vector notation:

$$FE_f(A) = (\nu(f([A]_1)) \ t \ \nu(f([A]_2)) \ t \ \ldots \ t \ \nu(f([A]_n)), \tag{3.52}$$
$$f([A]_1) \ t \ \nu(f([A]_2)) \ t \ \nu(f([A]_3)) \ t \ \ldots \ t \ \nu(f([A]_n)),$$
$$f([A]_1) \ t \ f([A]_2) \ t \ \nu(f([A]_3)) \ t \ \ldots \ t \ \nu(f([A]_n)),$$
$$\vdots$$
$$f([A]_1) \ t \ f([A]_2) \ t \ \ldots \ t \ f([A]_{n-1}) \ t \ \nu(f([A]_n)),$$
$$f([A]_1) \ t \ f([A]_2) \ t \ \ldots \ t \ f([A]_n)).$$

Because of the properties of the cardinality pattern f (Definition 3.7), and in particular its monotonicity, generalized f-weighted cardinalities of fuzzy sets keep all the properties of generalized cardinalities of fuzzy sets.

Example 3.3 For the fuzzy set (3.30) A from example (3.1):

$$A = 0.59/x_1 + 0.63/x_2 + 0.38/x_3 + 0.36/x_4 + 0.33/x_5 + 0.09/x_6. \qquad (3.53)$$

we received the following generalized cardinalities for specific cardinality patterns

- for cardinality pattern $f_{1,t}$ (3.25)

  - $FG_f(A) = (1, 1, 1, 0, 0, 0, 0, 0)$,
  - $FL_f(A) = (0, 0, 1, 1, 1, 1, 1, 1)$,
  - $FE_f(A) = (0, 0, 1, 0, 0, 0, 0, 0)$; $COG(FE_f(A)) = 2$.

  Their graphical interpretation has been presented in Fig. 3.7
- for cardinality pattern $f_{2,t,p}$ with parameters $t = 0.5$ and $p = 2$ (3.26)

  - $FG_f(A) = (1, 0.4, 0.35, 0, 0, 0, 0, 0,)$,
  - $FL_f(A) = (0.6, 0.65, 1, 1, 1, 1, 1, 1)$,
  - $FE_f(A) = (0, 0.4, 0.35, 0, 0, 0, 0, 0,)$; $COG(FE_f(A)) = 1.47$.

  Their graphical interpretation has been presented in Fig. 3.8
- for cardinality pattern $f_{3,t_1,t_2,p}$ (3.27) with parameters $t_1 = 0.3, t_2 = 0.7, p = 1$ we receive:

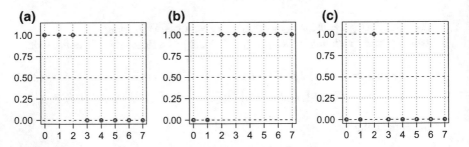

Fig. 3.7 Generalized f-weighted cardinal numbers of set A (3.53) and cardinality pattern $f_{1,t}$ (3.25) with parameters $t = 0.5, p = 2$: **a** $FG_f(A)$, **b** $FL_f(A)$, **c** $FE_f(A)$

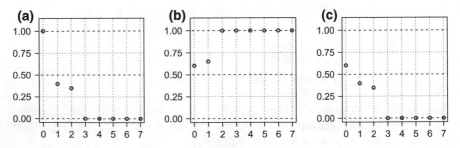

Fig. 3.8 Generalized f-weighted cardinal numbers of set A (3.53) and cardinality pattern $f_{2,t,p}$ (3.26) with parameters $t = 0.5$ and $p = 2$: **a** $FG_f(A)$, **b** $FL_f(A)$, **c** $FE_f(A)$

- $FG_f(A) = (1, 0.63, 0.59, 0.38, 0.36, 0.33, 0, 0)$,
- $FL_f(A) = (0.37, 0.41, 0.62, 0.64, 0.67, 1, 1, 1)$,
- $FE_f(A) = (0, 0.41, 0.59, 0.38, 0.36, 0.33, 0, 0)$; $COG(FE_f(A)) = 2.81$.

Their graphical interpretation has been presented in Fig. 3.9
- for cardinality pattern f_{4,t_1,t_2} (3.28) with parameters $t_1 = 0.3$, $t_2 = 0.7$

- $FG_f(A) = (1, 0.5, 0.5, 0.5, 0.5, 0.5, 0, 0,)$,
- $FL_f(A) = (0, 0.5, 0.5, 0.5, 0.5, 1, 1, 1)$,
- $FE_f(A) = (0, 0.5, 0.5, 0.5, 0.5, 0.5, 0, 0,)$; $COG(FE_f(A)) = 3$.

Their graphical interpretation has been presented in Fig. 3.10
- for cardinality pattern $f_{5,t,p}$ (3.29) with parameters $t = 0.5$, $p = 12$:

- $FG_f(A) = (1, 0.86, 0.78, 0.16, 0.12, 0.08, 0, 0)$,
- $FL_f(A) = (0.14, 0.22, 0.84, 0.88, 0.92, 1, 1, 1)$,
- $FE_f(A) = (0, 0.22, 0.78, 0.16, 0.12, 0.08, 0, 0)$; $COG(FE_f(A)) = 2.33$.

Their graphical interpretation has been presented in Fig. 3.11

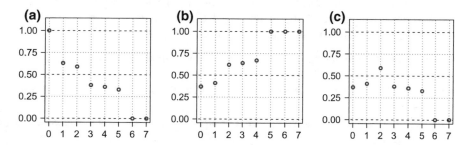

Fig. 3.9 Generalized f-weighted cardinal numbers of set A (3.53) and cardinality pattern $f_{3,t_1,t_2,p}$ (3.27) with parameters $t_1 = 0.3$, $t_2 = 0.7$, $p = 1$: **a** $FG_f(A)$, **b** $FL_f(A)$, **c** $FE_f(A)$

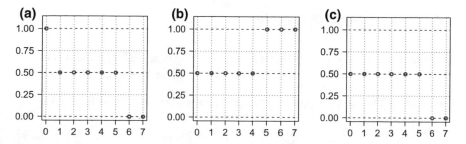

Fig. 3.10 Generalized f-weighted cardinal numbers of set A (3.53) and cardinality pattern f_{4,t_1,t_2} (3.28) with parameters $t_1 = 0.3$, $t_2 = 0.7$: **a** $FG_f(A)$, **b** $FL_f(A)$, **c** $FE_f(A)$

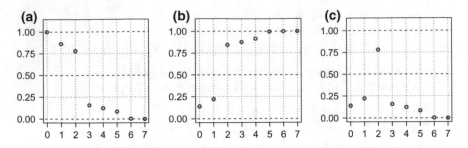

Fig. 3.11 Generalized f-weighted cardinal numbers of set A (3.53) and cardinality pattern $f_{5,t,p}$ (3.29) with parameters $t = 0.5$, $p = 12$: **a** $FG_f(A)$, **b** $FL_f(A)$, **c** $FE_f(A)$

3.5 Data Aggregation

The ability to aggregate information constitutes yet another very important element of data processing. This means the ability to represent sometime very big volumes of data with a single value expressing a certain aspect of information. Often these can be pieces of information from different sources of various types. Aggregation methods are of fundamental significance in decision making systems (see Dubois and Prade [34]) and their main areas of application include:

- they allow for comparing vectors or multi-dimensional matrices by comparing their representatives,
- they reduce the amount of information processes by substituting many pieces of it with just one,
- they allow to get some knowledge on a group of objects e.g. group of customers or market segment,
- they reduce the influence of atypical data (outliers) e.g. measurement errors.

3.5.1 Aggregation Operators

Let us start with formal, axiomatic formulation of a definition of aggregation operator that would allow us to determine the main features that an aggregator should exhibit.

Definition 3.13 An aggregation operator is an n-argument ($n \geq 1$) function

$$\text{Aggr} : \bigcup_{n \geq 1} [0, 1]^n \to [0, 1] \tag{3.54}$$

that meets the following conditions:

(A1) $\text{Aggr}(\underbrace{0, \ldots, 0)) = 0}_{n}$ and $\text{Aggr}(\underbrace{1, \ldots, 1) = 1}_{n}$,

(boundary conditions)

(A2) $\text{Aggr}(a_1, \ldots, a_n) \leq \text{Aggr}(b_1, \ldots, b_n)$ for $a_i \leq b_i$, $i = 1, \ldots, n$,

(monotonicity)

(A3) $\text{Aggr}(a) = a$ for each $a \in [0, 1]$,   (identity condition)

The number $\text{Aggr}(a_1, \ldots, a_n)$ is called aggregation of elements $a_1, \ldots, a_n$.

Aggregation operators Aggr can be divided into four classes (see Beliakov [10], Wygralak [150]):

1. Averaging operators—if they meet the condition:

$$(a_1 \wedge \ldots \wedge a_n) \leq \text{Aggr}(a_1, \ldots, a_n) \leq (a_1 \vee \ldots \vee a_n), \qquad (3.55)$$

2. Conjunctive operators—if they meet the condition:

$$\text{Aggr}(a_1, \ldots, a_n) \leq (a_1 \wedge \ldots \wedge a_n), \qquad (3.56)$$

3. Disjunctive operators—if they meet the condition:

$$\text{Aggr}(a_1, \ldots, a_n) \geq (a_1 \vee \ldots \vee a_n), \qquad (3.57)$$

4. Mixed operators—it means operators not belonging to any of the above-mentioned classes.

In forthcoming sections we will present examples of specific types of aggregation operators.

3.5.1.1 Minimum and Maximum

The minimum and maximum functions are a special type of functions that are aggregation operators of conjunctive and disjunctive type respectively, and in some limited cases also of averaging type.

We define the min ($\wedge$) and max ($\vee$) functions for $x \in [0, 1]^n$ any number of arguments $n \leq 1$ as:

$$\text{Aggr}_{\min}(x) = x_1 \wedge \ldots \wedge x_n \qquad (3.58)$$

$$\text{Aggr}_{\max}(x) = x_1 \vee \ldots \vee x_n \qquad (3.59)$$

3.5.1.2 Means

Means are aggregating functions of the averaging type. Basic and most frequently
used means include:

1. Arithmetic mean

$$\text{Aggr}_A(x) = \frac{1}{n} \sum_{i=1}^{n} x_i; \tag{3.60}$$

2. Geometric mean

$$\text{Aggr}_G(x) = \left(\prod_{i=1}^{n} x_i \right)^{1/n}; \tag{3.61}$$

3. Harmonic mean

$$\text{Aggr}_H(x) = n \left(\sum_{i=1}^{n} \frac{1}{x_i} \right)^{-1}; \tag{3.62}$$

4. r-mean

$$\text{Aggr}_R(x) = \left(\frac{1}{n} \sum_{i=1}^{n} x_i^r \right)^{\frac{1}{r}} \tag{3.63}$$

for any $r \in \mathbb{R}$. Arithmetic mean $Aggr_A$ is a special case of r-mean for $r = 1$.

3.5.1.3 Weighted Means

For a given vector $w = (w_1, \ldots, w_n)$ (i.e. weighting vector) for which $w_i > 0$ for
each i and $\sum_{i=1}^{n} w_i = 1$, we define:

1. Weighted arithmetic mean

$$\text{Aggr}_{WA}(x, w) = \sum_{i=1}^{n} w_i x_i; \tag{3.64}$$

2. Weighted geometric mean

$$\text{Aggr}_{WG}(x, w) = \prod_{i=1}^{n} x_i^{w_i}; \tag{3.65}$$

3. Weighted harmonic mean

$$\text{Aggr}_{\text{WH}}(x, w) = \left(\sum_{i=1}^{n} \frac{w_i}{x_i} \right)^{-1} ; \tag{3.66}$$

4. Weighted r-mean

$$\text{Aggr}_{\text{WR}}(x, w) = \left(\sum_{i=1}^{n} w_i x_i^r \right)^{\frac{1}{r}} \tag{3.67}$$

for any $r \in \mathbb{R}$. Arithmetic mean $Aggr_A$ is a special case of r-mean for $r = 1$.

3.5.1.4 OWA Operators (ordered Weighted Averaging)

OWA operators belong to the averaging function family. Their concept consist in assigning weights not to specific attributes, but to their values. OWA operators have been introduced by Yager [153] and because of their properties they soon became one of the most frequently used aggregation operators, among others in multiple-criteria decision-making.

Let $d_i, \ldots, d_n$ be a sequence obtained from $x_1, \ldots, x_n$ by ordering it in a non-increasing order, i.e. $d_1 \geq \ldots \geq d_n$, and let $w = (w_1, \ldots, w_n)$ be a (normalized) weighting vector with $w_i \geq 0$ for each $i = 1, \ldots, n$ and $\sum_{i=1}^{n} w_i = 1$. The OWA operator is defines as:

$$Aggr_{OWA}(x, w) = \sum_{i=1}^{n} w_i d_i; \tag{3.68}$$

Please note that if all the weights are equal, then the OWA operator is identical to arithmetic mean. If the vector $w = (1, 0, \ldots, 0)$ we receive maximum operator, and if the vector $w = (0, \ldots, 0, 1)$ we receive minimum operator.

3.5.1.5 Choquet and Sugeno Integrals

Both classes are averaging operators defined with a specific fuzzy measure. They are useful if we want to take into account interactions between variables x_i.

We call discrete fuzzy measure a function $\mu : 2^{\mathcal{N}} \to [0, 1]$ for $\mathcal{N} = \{1, 2, \ldots, n\}$ that is monotonous (i.e. $\mu(S) \leq \mu(T)$ if $S \subseteq T$) and meets the conditions $\mu(\varnothing) = 0$, $\mu(\mathcal{N}) = 1$

Discrete Choquet integral with respect to fuzzy measure μ is defined as:

$$\text{Agg}_{\text{Cho}}(x) = \sum_{i=1}^{n} \left[x_{(i)} - x_{(i-1)} \right] \mu(H_i) \qquad (3.69)$$

where $X_\pi = (x_{(1)}, x_{(2)}, \ldots, x_{(n)})$ is a non-decreasing permutation of input x. We assume that $x_{(0)} = 0$ and $H_i = \{(i), \ldots, (n)\}$ is a subset of indices $n - 1 + 1$ of the greatest element from x.

Special case of Choquet integrals include e.g. arithmetic mean, weighted mean and OWA.

Sugeno integral with respect to a fuzzy measure μ is defined as

$$\text{Agg}_{\text{Sug}}(x) = \max_{i=1,\ldots,n} \left[\min(\mu(H_i), x_{(i)}) \right] \qquad (3.70)$$

where $X_\pi = (x_{(1)}, x_{(2)}, \ldots, x_{(n)})$ is a non-decreasing permutation of input x, and $H_i = \{(i), \ldots, (n)\}$ is a subset of indices $n - 1 + 1$ of the greatest element from x.

3.5.1.6 Soft T-norms and T-conorms

Soft t-norms and t-conorms belong to the class of mixed operators. We define them in the following manner

$$\text{Agg}_\Phi(x, \alpha) = \frac{1 - \alpha}{n} \sum_{i=1}^{n} x_i + \alpha \cdot \Phi(x_1, \ldots, x_n) . \qquad (3.71)$$

where Φ is any t-norm or t-conorm and $\alpha \in [0, 1]$.

More on aggregation operators, including a detailed analysis of their properties and hints for using them in practice can be found in the monograph Beliakov et al. [10].

Chapter 4
Cardinalities of Interval-Valued Fuzzy Sets and Their Applications in Decision Making with Imperfect Information

Counting belongs to the most basic and frequent mental activities of humans. This is hardly surprising as cardinality seems to be one of the most fundamental characteristics of a given collection of elements. It forms an important type of information about that collection, a basis for coming to a decision in a lot of situations and in many dimensions of our life.

Wygralak [150]

In this chapter we introduce not fully determined fuzzy sets i.e. interval-valued fuzzy sets and Atanassov's intuitionistic fuzzy sets. We defined notions and methods describing cardinalities of such sets together with multiple examples. We presented methods for their application in decision-making algorithms with uncertain information. The chapter finishes with analysis of algorithms efficacy in ovarian cancer differentiation.

4.1 Decision Making with Imperfect Data

In many situations a decision maker faces the problems of data imperfect information on the basis of which he or she needs to make a decision. In such a case, one of the possible approaches is to use a couple of decision models operating on subsets of these data (often on non-disjoint subsets). Such decision models also return uncertain decisions, but they are often complementary and it is possible to aggregate them in an intelligent manner to make a decision far better than any of them made separately.

Formally speaking, we are dealing with a situation in which decisions come from independent decision sources (experts, systems). Each source expresses conviction level for a concrete decision. This conviction level is represented by a real value from the interval [0, 1], where 0 denotes sureness in negative decision (decision against a given option), and 1 designates full support for a positive decision. All the intermediate values mean that decision is burdened with uncertainty e.g. caused by incompleteness of input data. In particular, the value 0.5 means that decision

© Springer International Publishing AG 2018
K. Dyczkowski, *Intelligent Medical Decision Support System Based on Imperfect Information*, Studies in Computational Intelligence 735,
https://doi.org/10.1007/978-3-319-67005-8_4

source is unable to make any decision as there are as many arguments for as there are against. Additionally, if decision source had gaps in input data necessary for decision making, the output decision may then be represented by an interval. In other words, this interval designates upper and lower limit for possible decisions for this source. Please note, that such an interval represents epistemic knowledge i.e. it includes the exact value of decision that the source would take if it had full knowledge needed for it (if there were no gaps in input data). The task is thus to intelligently aggregate many decisions coming from representative source in the form of intervals. A skillful fusion of such decision sources should allow for making accurate and high-quality decisions even with data imperfect input data.

4.2   Interval-Valued Fuzzy Sets

Interval-valued fuzzy sets (IVFSs) have been defined in 1970s almost simultaneously by Zadeh (see Zadeh [155]) and Grattan-Guiness (see Grattan-Guiness [49]). This was a response to criticism of fuzzy sets relating to them not representing lack of precision. The critics argued that fuzzy sets only grade memberships and they are unable of expressing gaps in knowledge. Interval-valued fuzzy sets are a natural extension of fuzzy sets where instead of a single value of membership function each element is associated with an interval. This allows for expansion of the fuzzy set theory and using them to represent gradability of membership taking into account uncertainty and gaps in data.

Interval-valued fuzzy sets are a special variant of type-2 fuzzy sets, also introduced by Zadeh (see Zadeh [155]). Walker and Walker [143] is also worth a look at, as it describes precisely operations on type-2 fuzzy sets. Results described therein also apply to special cases of IVFS. In Mendel and John [77], Mendel et al. [78] theoretical and practical aspects of such objects have been described. Please note that if such objects do not suffer from lack of precisions, they are reduced to regular type-1 fuzzy sets.

What is worth stressing here is the modelling idea for incomplete known fuzzy sets as introduced by Wygralak [150]. It is a general concept of object that are capable of modelling notions with gaps in knowledge on element membership. This concept lies in representation of set $A \in U$ with the use of three disjoint classes:

- A^+—set composed of elements that **for sure belong** to set A;
- A^-—set composed of elements that **for sure do not belong** to set A;
- $A^?$—set of elements with respect to which **we are unsure** or **we have no knowledge**, as to whether they belong to set A (the so-called uncertainty region).

From this definition of class division we conclude that A is somewhere in between A^+ and $(A^-)'$:

$$A^+ \subset A \subset (A^-)' \text{ and } A^- \subset A' \subset (A^+)' \tag{4.1}$$

Moreover, A^+ is the lower bound of A, and A^- is the lower bound of A' and:

$$A^+ \cap A^? = (A^-)' \quad \text{and} \quad A^? = (A^+ \cap A^-)'. \tag{4.2}$$

Each of the three sets A^+, A^-, $A^?$ is defined in a unique manner by the two others. Thus, when defining them it is enough to specify any pair.

Interval-valued fuzzy sets can be well placed in this concept with uncertainty represented by a pair $(A^+, A^+ \cap A^?)$ thus a set of such elements that exhibit a given feature and a sum of sets not containing it or not sure of it.

Let us start with defining the notion of IVFS.

Definition 4.1 Let $\mathcal{U}$ be a universe of arbitrary elements, finite or not. We call interval-valued fuzzy set in $\mathcal{U}$ an object identified with function

$$\widetilde{A} : \mathcal{U} \to \mathcal{I}([0, 1]) \tag{4.3}$$

where $\mathcal{I}([0, 1])$ is a set of all closed intervals included in $[0, 1]$ i.e. $\mathcal{I}([0, 1]) = \{[a, b] : a, b \in [0, 1]\}$. Function $\widetilde{A}$ is called membership function of IVFS $\widetilde{A}$.

In further analysis interval-valued fuzzy sets will be denoted with a tilde to differentiate them from usual fuzzy sets: $\widetilde{A}$, $\widetilde{B}$ etc.

The notion of interval-valued fuzzy sets is a generalization of the notion of an usual fuzzy set. Its significant role is to introduce uncertainty as to an actual value of membership function (epistemic interpretation of interval-valued fuzzy set (see Dubois and Prade [35])) that can be anywhere between the given interval boundaries.

Interval-valued fuzzy set can also be defined as a pair of two fuzzy sets A_l, A_u defined in the universe $\mathcal{U}$, where A_l and A_u respectively, denote a lower and upper bound of the membership interval:

$$\widetilde{A} = (A_l, A_u) \tag{4.4}$$

such that for each $x \in \mathcal{U}$, $A_l(x) \le A_u(x)$ (fuzzy set A_l is included in A_u i.e. $A_l \subseteq A_u$).

Such a construction of IVFS is most frequently used for applications because of the ease of operating it with notions pertaining to regular fuzzy sets. In Fig. 4.1 we presented an example of interval-fuzzy set written in singleton form:

$$\widetilde{A} = \frac{(0.2, 0.67)}{x_1} + \frac{(0.07, 0.67)}{x_2} + \frac{(0.06, 0.25)}{x_3} + \frac{(0.02, 0.023)}{x_4} + \\ + \frac{(0, 0.83)}{x_5} + \frac{(0, 0.05)}{x_6}. \tag{4.5}$$

The same fuzzy set can be represented as a pair of fuzzy sets $\widetilde{A} = (A_l, A_u)$:

$$A_l = \frac{0.2}{x_1} + \frac{0.07}{x_2} + \frac{0.06}{x_3} + \frac{0.02}{x_4} + \frac{0}{x_5} + \frac{0}{x_6}, \\ A_u = \frac{0.67}{x_1} + \frac{0.67}{x_2} + \frac{0.25}{x_3} + \frac{0.023}{x_4} + \frac{0.83}{x_5} + \frac{0.05}{x_6}. \tag{4.6}$$

Fig. 4.1 Interval-valued
fuzzy set $\widetilde{A}$ (4.5)

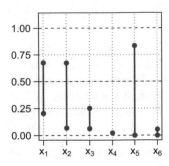

4.2.1 Operations on Interval-Valued Fuzzy Sets

Union, product and complement of interval-valued fuzzy sets (IVFS) are obtained
by generalizing definitions of these operations on regular fuzzy sets. Let us use
representation (4.6) of interval-valued fuzzy set based on two fuzzy sets A_l, A_u.

Let $\widetilde{A} = (A_l, A_u)$ and $\widetilde{B} = (B_l, B_u)$ be two interval-valued fuzzy sets (IVFS), let
v denote a strong negation and let t, s denote a t-norm and a t-conorm respectively.
Basic operations on IVFSs can be defined as follows:

$$\widetilde{A} = \widetilde{B} \iff A_l = B_l \text{ and } A_u = B_{u(\text{identity})}, \tag{4.7}$$

$$\widetilde{A} \subset \widetilde{B} \iff A_l \subset B_l \text{ and } A_u \subset B_{u(\text{inclusion})}, \tag{4.8}$$

$$\widetilde{A} \cap_t \widetilde{B} = (A_l \cap_t B_l, \ A_u \cap_t B_u)_{(\text{intersection})}, \tag{4.9}$$

$$\widetilde{A} \cup_s \widetilde{B} = (A_l \cup_s B_l, \ A_u \cup_s B_u)_{(\text{union})}, \tag{4.10}$$

$$\left(\widetilde{A}\right)^v = (A_u^v, \ A_l^v)_{(\text{complement})}. \tag{4.11}$$

If using standard operations i.e. $t = \wedge$, $s = \vee$ and $v = v_L$ we can use a
simplified notation i.e. $\widetilde{A} \cap \widetilde{B}, \widetilde{A} \cup \widetilde{B}, \widetilde{A}'$.

More information on other ways of defining operations on IFVSs can be found in
Wygralak [150], Bustince et al. [15], Deschrijver et al. [29].

4.3 Atanassov's Intuitionistic Fuzzy Sets

Another approach to generalize the notion of fuzzy sets with the uncertainty modelling are intuitionistic fuzzy sets proposed in 1980s by Atanassov (see Atanassov
[4, 5]). The idea behind this concept was to define such fuzzy sets as a pair of membership functions simultaneously describing membership and non-membership of a
given element and assumption that the sum of both values of these functions does

not exceed 1. Calling these sets "intuitionistic" caused heated discussion in scientific community (see Dubois et al. [31]), thus such sets are often called IF-sets (IFSs) for short.

Definition 4.2 Let $\mathcal{U}$ be a universe of arbitrary elements, finite or not. We call an intuitionistic fuzzy set $\mathcal{A}$ in $\mathcal{U}$ a pair of fuzzy sets (A^+, A^-) such that

$$A^+ \subset (A^-)^\nu. \tag{4.12}$$

where ν is a strong negation.

We call A^+ membership function and A^- non-membership function.

The idea behind them corresponds to the concept of fuzzy sets not fully known described in the previous section. In such a case, information on positive membership A^+ and negative membership (non-membership) A^- has been used to describe uncertainty.

For example, IVFS $\widetilde{A}$ with (4.6) can be represented as IFS $\mathcal{A} = (A^+, A^-)$:

$$
\begin{aligned}
A^+ &= \frac{0.2}{x_1} + \frac{0.07}{x_2} + \frac{0.06}{x_3} + \frac{0.02}{x_4} + \frac{0}{x_5} + \frac{0}{x_6} \\
A^- &= \frac{0.23}{x_1} + \frac{(0.33)}{x_2} + \frac{0.75}{x_3} + \frac{0.977}{x_4} + \frac{0.17}{x_5} + \frac{0.95}{x_6}.
\end{aligned}
\tag{4.13}
$$

In Fig. 4.2 we presented its graphical representation (to stress bipolar character of information presentation, both components have been presented in the same graph, rotated symmetrically towards one another).

Please note that taking into account (4.12) IVFSs and IFSs are formally equivalent notions i.e. (A^+, A^-) is IFS iff $(A^+, (A^-)^\nu)$ is IVFS. But taking into account practical applications these two notions vary as IFS introduce bipolar perspective putting some stress on positive and negative information. Such a perspective is very often useful in presenting uncertain information to a human being. It stresses arguments for and against and presents the scope of expert's lack of knowledge in an intuitionistic manner (see e.g. Dubois and Prade [35], Stachowiak et al. [113]).

Another important aspect in modelling IFS is the notion of hesitation margin that represents information on the degree to which we are lacking information on

Fig. 4.2 Atanassov's intuitionistic fuzzy set $\mathcal{A}$ (4.13)

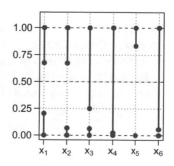

positive and negative features of the object/decision under consideration. It mirrors the notions of "neither for, nor against". It can be generally defined as (cf. Wygralak [150]):

$$\chi_{\mathcal{A}} = (A^+)^\nu \cap_t (A^-)^\nu = (A^+ \cup_{t^\nu} A^-)^\nu. \tag{4.14}$$

Please note that it is a general formulation for any strong negation ν and t-norm t generated by it. If we are using Łukasiewicz negation $\nu = \nu_L$ and Łukasiewicz t-norm $t = t_L$ we obtain original definitions of hesitation margin as proposed by Atanassov (see Atanassov [5]) i.e. for each $x \in \mathcal{U}$:

$$\chi_{\mathcal{A}}(x) = 1 - A^+(x) - A^-(x), \tag{4.15}$$

where the formula (4.12) can be reduced to the condition:

$$A^+(x) + A^-(x) \le 1, \tag{4.16}$$

and

$$A^+(x) + A^-(x) + \chi_{\mathcal{A}}(x) = 1. \tag{4.17}$$

A very comprehensive monographic take on IFS, with special attention paid to probability measures can be found in Szmidt [118] and the latest monograph Atanassov [6]. More considerations on hesitation margin, IFSs cardinalities and their applications can be found in Pankowska and Wygralak [97, 98], Wygralak [150].

Further in this book, we will focus on interval-valued fuzzy sets and because of their mathematical identity with IFSs we will not introduce separate versions for the latter ones. However, in the part devoted to diagnostic systems they have been used for more intuitive presentation of current medical diagnosis.

In this place, it is worth mentioning yet other extension of fuzzy sets i.e. ordered fuzzy numbers, introduced by Kosiński et al. [71] which has many practical applications (see e.g. Czerniak et al. [20–22]).

4.4 Scalar Cardinality of IVFSs (*Sigma F-Counts*)

In this part we will introduce a definition of scalar cardinalities of interval-valued fuzzy sets. Cardinalities of interval-valued fuzzy sets are defined in a natural manner using scalar cardinalities of fuzzy sets described in Sect. 3.4.1.

For a finite interval-valued fuzzy set $\widetilde{A} = (A_l, A_u)$ with $A_l \subset A_u$ scalar cardinality of type *sigma f-Count* $sc_f(\widetilde{A})$ is defined as an interval of non-negative real numbers (see Wygralak [150]):

$$sc_f(\widetilde{A}) = [sc_f(A_l), sc_f(A_u)], \tag{4.18}$$

where f is the cardinality pattern introduced in Definition 3.8. Cardinality of IFSs is defined similarly. For IFS $\mathcal{A} = (A^+, A^-)$ where $A^+ \subset (A^-)^\nu$ with strong negation ν *sigma f-Count* $sc_f(\mathcal{A})$ is defined as

$$sc_f(\mathcal{A}) = \left[sc_f(A^+), sc_f((A^-)^\nu) \right]. \qquad (4.19)$$

An interval created in this way is interpreted as an interval of possible values of cardinalities of not fully determined fuzzy set with its left endpoint representing minimum possible scalar cardinality and the right endpoint representing the maximum possible scalar cardinality.

We say that two interval-valued fuzzy sets $\widetilde{A}$ and $\widetilde{B}$ are equipotent if $sc_f(\widetilde{A}) = sc_f(\widetilde{B})$. Equipotency is defined similarly for IFSs.

Because, as we have mentioned before, IVFS and IFS are formally equivalent notions, further in this book we will focus on theory elements and examples for IVFS. It is easy to introduce their equivalents for IFSs.

If, in a concrete application it is necessary to approximate cardinality of IVFS with a single real number (e.g. for the purpose of comparison of cardinalities of two IVFSs), the notion of cardinality representative *Rep* is used. Cardinality of interval-valued fuzzy set can be approximated with a couple of methods:

- Upper estimation (maximum possible cardinality):

$$Rep_{max} = max(sc_f(A_l), sc_f(A_u)) = sc_f(A_u), \qquad (4.20)$$

- Lower estimation (minimum possible cardinality):

$$Rep_{min} = min(sc_f(A_l), sc_f(A_u)) = sc_f(A_l), \qquad (4.21)$$

- Average estimation (mean possible cardinality):

$$Rep_{avg} = \left[sc_f(A_l) + sc_f(A_u) \right] / 2. \qquad (4.22)$$

Readers interested in cardinalities of IVFSs and IFS and its axiomatic may find more information in the following publications: Wygralak [149, 150], Deschrijver and Král [30], Král [72].

We will present an example how to determine cardinalities of interval-valued fuzzy sets with various cardinality patterns.

Example 4.1 IVFS $\widetilde{A} = (A_l, A_u)$ is defined in the following manner:

$$A_l = 0.42/x_1 + 0.27/x_2 + 0.75/x_3 + 0.16/x_4 + 0.33/x_5 + 0.52/x_6,$$
$$A_u = 0.73/x_1 + 0.75/x_2 + 0.75/x_3 + 0.51/x_4 + 0.50/x_5 + 0.52/x_6. \qquad (4.23)$$

Its graphical representation has been presented in Fig. 4.3.
Scalar cardinality for various cardinality patterns together with average cardinality estimation Rep_{avg} is the following:

Fig. 4.3 Graphical interpretation of IVFS $\widetilde{A}$ (4.23)

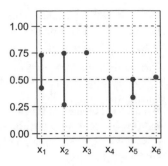

- For function $f = id$

$$sc_f(\widetilde{A}) = (2.45, 3.76); \quad Rep_{avg}(sc_f(\widetilde{A})) = 3.1,$$

- If using function $f = f_1$ (3.25) with parameters $t = 0.5$, $p = 2$ as cardinality pattern, we obtain the following cardinalities:

$$sc_f(\widetilde{A}) = (2, 5); \quad Rep_{avg}(sc_f(\widetilde{A})) = 3.5,$$

- If using function $f = f_2$ (3.26) with parameters $t = 0.5$, $p = 1$ as cardinality patterns, we obtain the following cardinalities:

$$sc_f(\widetilde{A}) = (1.27, 3.26); \quad Rep_{avg}(sc_f(\widetilde{A})) = 2.2,$$

- If using function $f = f_3$ (3.27) with parameters $t_1 = 0.4$, $t_2 = 0.6$, $p = 1$ as cardinality pattern, we obtain the following cardinalities:

$$sc_f(\widetilde{A}) = (1.94, 4.53); \quad Rep_{avg}(sc_f(\widetilde{A})) = 3.2,$$

- If using function $f = f_4$ (3.28) with parameters $t_1 = 0.4$, $t_2 = 0.6$, $p = 1$ as cardinality pattern, we obtain the following cardinalities:

$$sc_f(\widetilde{A}) = (2, 4.5); \quad Rep_{avg}(sc_f(\widetilde{A})) = 3.25,$$

- If using function $f = f_5$ (3.29) with parameters $t = 0.5$, $p = 12$ as cardinality pattern, we obtain the following cardinalities:

$$sc_f(\widetilde{A}) = (1.92, 4.51); \quad Rep_{avg}(sc_f(\widetilde{A})) = 3.21.$$

Because of its simplicity and low computational complexity, this method of calculating cardinalities of interval-valued fuzzy sets is the most frequently used one. However, in more demanding applications, where very precise description of cardinality is needed, it may prove insufficient.

4.5 Fuzzy Cardinality of IVFSs (*f-FCounts*)

Cardinalities of interval-valued fuzzy sets are defined in a natural manner using cardinalities of fuzzy sets described in Sect. 3.4.2.

For a finite interval-valued fuzzy set $\widetilde{A} = (A_l, A_u)$ with $A_l \subseteq A_u$ fuzzy type cardinalities are defined as interval-valued fuzzy sets in $\mathbb{N}$ (see Wygralak [148]).

Definition 4.3 1. *f-FGCount* of IVFS $\widetilde{A}$ for a given cardinality pattern f is defined as:

$$FG_f(\widetilde{A}) = (FG_f(A_l), FG_f(A_u)), \tag{4.24}$$

i.e. for $k \in \mathbb{N}$:

$$\begin{aligned} FG_f(\widetilde{A})(k) &= [FG_f(A_l)(k), FG_f(A_u)(k)] = \\ &= [f([A_l]_k), f([A_u]_k)], \end{aligned} \tag{4.25}$$

where $FG_f(A_l)$ and $FG_f(A_u)$ are the fuzzy cardinalites defined in (3.48).

2. *f-FLCount* of IVFS $\widetilde{A}$ for a given cardinality pattern f is defined as:

$$FL_f(\widetilde{A}) = (FL_f(A_u), FL_f(A_l)) \tag{4.26}$$

i.e. for $k \in \mathbb{N}$:

$$\begin{aligned} FL_f(\widetilde{A})(k) &= [FL_f(A_l)(k), FL_f(A_u)(k)] = \\ &= [1 - f([A_u]_{k+1}), 1 - f([A_l]_k)], \end{aligned} \tag{4.27}$$

where $FL_f(A_l)$ and $FL_f(A_u)$ are the fuzzy cardinalites defined in (3.50).

3. *f-FECount* of IVFS $\widetilde{A}$ for a given cardinality pattern f is defined as:

$$FE_f(\widetilde{A}) = FG_f(\widetilde{A}) \cap FL_f(\widetilde{A}) \tag{4.28}$$

i.e. for $k \in \mathbb{N}$:

$$\begin{aligned} FE_f(\widetilde{A})(k) &= [f([A_l]_k) \wedge (1 - f([A_u]_{k+1})), \\ & \quad f([A_u]_k) \wedge (1 - f([A_l]_{k+1})]. \end{aligned} \tag{4.29}$$

At the beginning, it is worth presenting some examples for special cases of interval-valued fuzzy sets (with the use of cardinality pattern $f = id$).

Example 4.2 For IVFS $\widetilde{A} = (A_l, A_u)$ with

$$\begin{aligned} A_l &= 1/x_1 + 1/x_2 + 1/x_3 + 1/x_4 + 1/x_5 + 1/x_6, \\ A_u &= 1/x_1 + 1/x_2 + 1/x_3 + 1/x_4 + 1/x_5 + 1/x_6, \end{aligned} \tag{4.30}$$

whose graphical representation has been presented in Fig. 4.4 i.e. a crisp set equal to 1 over the whole support, values of fuzzy cardinalities *f-Fcounts* are the following

Fig. 4.4 Graphical interpretation of IVFS $\widetilde{A}$ (4.30)

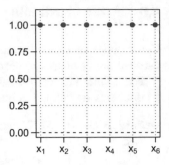

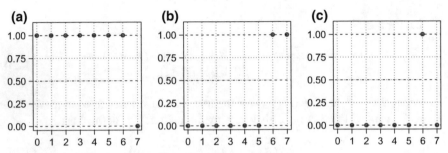

Fig. 4.5 Fuzzy cardinalities of IVFS $\widetilde{A}$ (3.53) for cardinality pattern $f = f_{id}$: **a** $FG_f(\widetilde{A})$, **b** $FL_f(\widetilde{A})$, **c** $FE_f(\widetilde{A})$

$$
\begin{aligned}
FG_f(\widetilde{A}) &= ((1, 1, 1, 1, 1, 1, 1, 0), \\
&\quad\;\; (1, 1, 1, 1, 1, 1, 1, 0)); \\
FL_f(\widetilde{A}) &= ((0, 0, 0, 0, 0, 0, 1, 1), \\
&\quad\;\; (0, 0, 0, 0, 0, 0, 1, 1)); \\
FE_f(\widetilde{A}) &= ((0, 0, 0, 0, 0, 0, 1, 0), \\
&\quad\;\; (0, 0, 0, 0, 0, 0, 1, 0));
\end{aligned}
\tag{4.31}
$$

One can see that the value of cardinality corresponds to our intuition i.e. cardinality of a crisp set with at least, at most and exactly 6 elements. Graphical interpretation of IVFS cardinality has been presented in Fig. 4.5

Example 4.3 For IVFS $\widetilde{A} = (A_l, A_u)$:

$$
\begin{aligned}
A_l &= 0/x_1 + 0/x_2 + 0/x_3 + 0/x_4 + 0/x_5 + 0/x_6, \\
A_u &= 0/x_1 + 0/x_2 + 0/x_3 + 0/x_4 + 0/x_5 + 0/x_6.
\end{aligned}
\tag{4.32}
$$

whose graphical representation has been presented in Fig. 4.6 i.e. for a crisp set equal to 0 over the whole support, values of fuzzy cardinalities *f-Fcounts* are the following

Fig. 4.6 Graphical interpretation of IVFS $\widetilde{A}$ (4.32)

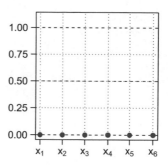

$$FG_f(\widetilde{A}) = ((1, 0, 0, 0, 0, 0, 0, 0),$$
$$(1, 0, 0, 0, 0, 0, 0, 0));$$
$$FL_f(\widetilde{A}) = ((1, 1, 1, 1, 1, 1, 1, 1),$$
$$(1, 1, 1, 1, 1, 1, 1, 1));$$
$$f(\widetilde{A}) = ((1, 0, 0, 0, 0, 0, 0, 0),$$
$$(1, 0, 0, 0, 0, 0, 0, 0));$$

(4.33)

One can see that fuzzy cardinalities of IVFS $\widetilde{A}$ correspond to intuitive cardinality of a crisp set containing at least, at most and exactly zero elements. Their graphical interpretation has been presented in Fig. 4.7

Example 4.4 Let us now consider a situation in which IVFS includes complete information (it constitutes a fuzzy set), but there is full uncertainty as to membership of each of elements i.e. set $\widetilde{A} = (A_l, A_u)$:

$$A_l = 0.5/x_1 + 0.5/x_2 + 0.5/x_3 + 0.5/x_4 + 0.5/x_5 + 0.5/x_6,$$
$$A_u = 0.5/x_1 + 0.5/x_2 + 0.5/x_3 + 0.5/x_4 + 0.5/x_5 + 0.5/x_6,$$

(4.34)

whose graphical representation has been presented in Fig. 4.8. When using identity function ($f = f_{id}$) as cardinality pattern we obtain the following cardinalities

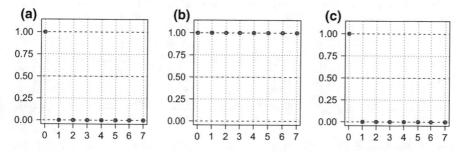

Fig. 4.7 Fuzzy cardinalites of IVFS $\widetilde{A}$ (3.53) with cardinality pattern $f = f_{id}$: **a** $FG_f(\widetilde{A})$, **b** $FL_f(\widetilde{A})$, **c** $FE_f(\widetilde{A})$

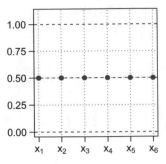

$$FG_f(\widetilde{A}) = ((1, 0.5, 0.5, 0.5, 0.5, 0.5, 0.5, 0),$$
$$(1, 0.5, 0.5, 0.5, 0.5, 0.5, 0.5, 0));$$
$$FL_f(\widetilde{A}) = ((0.5, 0.5, 0.5, 0.5, 0.5, 0.5, 1, 1),$$
$$(0.5, 0.5, 0.5, 0.5, 0.5, 0.5, 1, 1));$$ \hfill (4.35)
$$FE_f(\widetilde{A}) = ((0.5, 0.5, 0.5, 0.5, 0.5, 0.5, 0.5, 0),$$
$$(0.5, 0.5, 0.5, 0.5, 0.5, 0.5, 0.5, 0));$$

It can be seen in this case that there is full hesitation as to whether cardinality of such
a set is equal to one of the values from set $\{1, 2, 3, 4, 5, 6\}$. Graphical interpretation
of fuzzy cardinalities of IVFS $\widetilde{A}$ has been presented in Fig. 4.9.

Example 4.5 As the last example of a special case let us consider a situation in
which an IVFS has fully incomplete information i.e. IVFS $\widetilde{A} = (A_l, A_u)$:

$$A_l = 0/x_1 + 0/x_2 + 0/x_3 + 0/x_4 + 0/x_5 + 0/x_6,$$
$$A_u = 1/x_1 + 1/x_2 + 1/x_3 + 1/x_4 + 1/x_5 + 1/x_6.$$ \hfill (4.36)

Its graphical representation has been presented in Fig. 4.10. When using identity
function ($f = f_{id}$) as cardinality pattern we obtain the following cardinalities

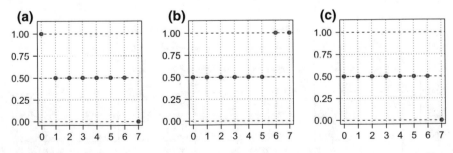

Fig. 4.9 Fuzzy cardinalities of IVFS A (3.53) with cardinality pattern $f = f_{id}$: **a** $FG_f(\widetilde{A})$, **b**
$FL_f(\widetilde{A})$, **c** $FE_f(\widetilde{A})$

Fig. 4.10 Graphical interpretation of IVFS $\widetilde{A}$ (4.36)

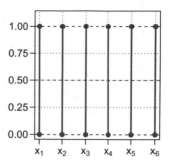

$$FG_f(\widetilde{A}) = ((1, 0, 0, 0, 0, 0, 0, 0),$$
$$(1, 1, 1, 1, 1, 1, 1, 0));$$
$$FL_f(\widetilde{A}) = ((0, 0, 0, 0, 0, 0, 1, 1),$$
$$(1, 1, 1, 1, 1, 1, 1, 1));$$
$$FE_F(\widetilde{A}) = ((0, 0, 0, 0, 0, 0, 0, 0),$$
$$(1, 1, 1, 1, 1, 1, 1, 0)).$$

(4.37)

In this case fuzzy cardinalities of IVFS $\widetilde{A}$ mirror lack of information (interval [0, 1]) on membership to specific cardinality values. Their graphical interpretation has been presented in Fig. 4.11.

Example 4.6 Let us now consider the following IVFS $\widetilde{A} = (A_l, A_u)$:

$$A_l = 0.55/x_1 + 0.59/x_2 + 0.38/x_3 + 0.12/x_4 + 0.83/x_5 + 0/x_6,$$
$$A_u = 0.55/x_1 + 0.59/x_2 + 0.50/x_3 + 0.78/x_4 + 1/x_5 + 0.65/x_6.$$

(4.38)

Its graphical representation has been presented in Fig. 4.12. Let us now present fuzzy cardinalities of this IVFS for various cardinality patterns:

1. In the case of using identity function ($f = f_{id}$) as cardinality pattern we obtain the following fuzzy cardinalities of IVFS $\widetilde{A}$ (4.38):

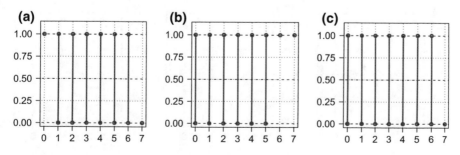

Fig. 4.11 Fuzzy cardinalities of IVFS $\widetilde{A}$ (3.53) with cardinality pattern $f = f_{id}$: **a** $FG_f(\widetilde{A})$, **b** $FL_f(\widetilde{A})$, **c** $FE_f(\widetilde{A})$

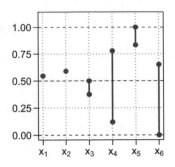

Fig. 4.12 Graphical interpretation of IVFS $\widetilde{A}$ (4.38)

$$
\begin{aligned}
FG_f(\widetilde{A}) = &((1, 1, 0.78, 0.65, 0.59, 0.55, 0.50, 0), \\
&(1, 0.83, 0.59, 0.55, 0.38, 0.12, 0, 0)); \\
FL_f(\widetilde{A}) = &((0, 0.22, 0.35, 0.41, 0.45, 0.5, 1, 1), \\
&(0.17, 0.41, 0.45, 0.62, 0.88, 1, 1, 1)); \\
FE_f(\widetilde{A}) = &((0, 0.22, 0.35, 0.41, 0.38, 0.12, 0, 0), \\
&(0.17, 0.41, 0.45, 0.62, 0.59, 0.55, 0.5, 0)).
\end{aligned}
\tag{4.39}
$$

Their graphical interpretation has been presented in Fig. 4.13.

2. If using function $f = f_1$ (3.25) with parameters $t = 0.5$, $p = 2$ as cardinality pattern we obtain the following cardinalities:

$$
\begin{aligned}
FG_f(\widetilde{A}) = &((1, 1, 1, 1, 0, 0, 0, 0), \\
&(1, 1, 1, 1, 1, 1, 0, 0)); \\
FL_f(\widetilde{A}) = &((0, 0, 0, 0, 0, 1, 1, 1), \\
&(0, 0, 0, 1, 1, 1, 1, 1)); \\
FE_f(\widetilde{A}) = &((0, 0, 0, 0, 0, 0, 0, 0), \\
&(0, 0, 0, 1, 1, 1, 0, 0)).
\end{aligned}
\tag{4.40}
$$

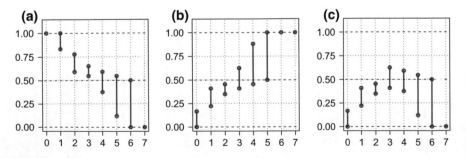

Fig. 4.13 Fuzzy cardinalities of IVFS $\widetilde{A}$ (4.38) with cardinality pattern $f = f_{id}$: **a** $FG_f(\widetilde{A})$, **b** $FL_f(\widetilde{A})$, **c** $FE_f(\widetilde{A})$

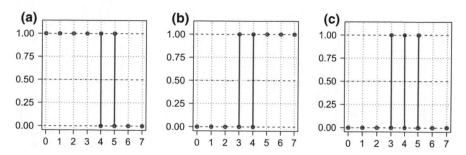

Fig. 4.14 Fuzzy cardinalities of IVFS $\widetilde{A}$ (4.38) with cardinality pattern $f = f_1$ (3.25): **a** $FG_f(\widetilde{A})$, **b** $FL_f(\widetilde{A})$, **c** $FE_f(\widetilde{A})$

Their graphical interpretation has been presented in Fig. 4.14.

3. If using function $f = f_2$ (3.26) with parameters $t = 0.5$, $p = 1$ as cardinality patterns we obtain the following cardinalities:

$$
\begin{aligned}
FG_f(\widetilde{A}) &= ((1, 0.69, 0.35, 0.3, 0, 0, 0, 0), \\
&\quad\ (1, 1, 0.61, 0.42, 0.35, 0.3, 0, 0)); \\
FL_f(\widetilde{A}) &= ((0, 0.39, 0.58, 0.65, 0.7, 1, 1, 1), \\
&\quad\ (0.31, 0.65, 0.7, 1, 1, 1, 1, 1)); \\
FE_f(\widetilde{A}) &= ((0, 0.39, 0.35, 0.3, 0, 0, 0, 0), \\
&\quad\ (0.31, 0.65, 0.61, 0.42, 0.35, 0.3, 0, 0)).
\end{aligned}
\tag{4.41}
$$

Their graphical interpretation has been presented in Fig. 4.15.

4. If using function $f = f_3$ (3.27) with parameters $t_1 = 0.4$, $t_2 = 0.6$, $p = 1$ as cardinality pattern we obtain the following cardinalities:

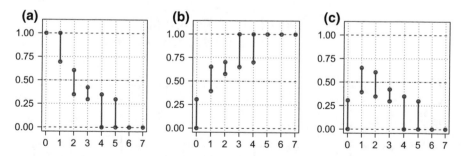

Fig. 4.15 Fuzzy cardinalities of IVFS $\widetilde{A}$ (4.38) with cardinality pattern $f = f_2$ (3.26): **a** $FG_f(\widetilde{A})$, **b** $FL_f(\widetilde{A})$, **c** $FE_f(\widetilde{A})$

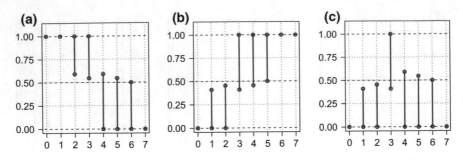

Fig. 4.16 Fuzzy cardinalities of IVFS $\widetilde{A}$ (4.38) with cardinality pattern $f = f_3$ (3.27): **a** $FG_f(\widetilde{A})$, **b** $FL_f(\widetilde{A})$, **c** $FE_f(\widetilde{A})$

$$
\begin{aligned}
FG_f(\widetilde{A}) &= ((1, 1, 0.59, 0.55, 0, 0, 0, 0,), \\
&\quad (1, 1, 1, 1, 0.59, 0.55, 0.5, 0)); \\
FL_f(\widetilde{A}) &= ((0, 0, 0, 0.41, 0.45, 0.5, 1, 1), \\
&\quad (0, 0.41, 0.45, 1, 1, 1, 1, 1)); \\
FE_f(\widetilde{A}) &= ((0, 0, 0, 0.41, 0, 0, 0, 0), \\
&\quad (0, 0.41, 0.45, 1, 0.59, 0.55, 0.5, 0)).
\end{aligned}
\tag{4.42}
$$

Their graphical interpretation has been presented in Fig. 4.16.

5. If using function $f = f_4$ (3.28) with parameters $t_1 = 0.4$, $t_2 = 0.6$, $p = 1$ as cardinality pattern we obtain the following cardinalities:

$$
\begin{aligned}
FG_f(\widetilde{A}) &= ((1, 1, 0.5, 0.5, 0, 0, 0, 0,), \\
&\quad (1, 1, 1, 1, 0.5, 0.5, 0.5, 0)); \\
FL_f(\widetilde{A}) &= ((0, 0, 0, 0.5, 0.5, 0.5, 1, 1), \\
&\quad (0, 0.5, 0.5, 1, 1, 1, 1, 1)); \\
FE_f(\widetilde{A}) &= ((0, 0, 0, 0.5, 0, 0, 0, 0), \\
&\quad (0, 0.5, 0.5, 1, 0.5, 0.5, 0.5, 0)).
\end{aligned}
\tag{4.43}
$$

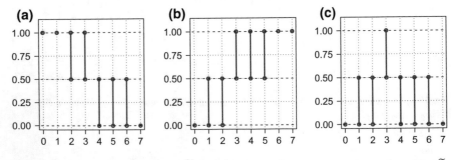

Fig. 4.17 Fuzzy cardinalities of IVFS $\widetilde{A}$ (4.38) with cardinality pattern $f = f_4$ (3.28): **a** $FG_f(\widetilde{A})$, **b** $FL_f(\widetilde{A})$, **c** $FE_f(\widetilde{A})$

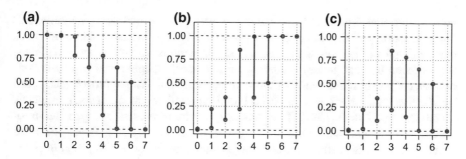

Fig. 4.18 Fuzzy cardinalities of IVFS $\widetilde{A}$ (4.38) with cardinality pattern $f = f_5$ (3.29): **a** $FG_f(\widetilde{A})$, **b** $FL_f(\widetilde{A})$, **c** $FE_f(\widetilde{A})$

Their graphical interpretation has been presented in Fig. 4.17.

6. If using function $f = f_5$ (3.29) with parameters $t = 0.5$, $p = 12$ as cardinality pattern we obtain the following cardinalities:

$$
\begin{aligned}
FG_f(\widetilde{A}) &= ((1, 0.99, 0.78, 0.66, 0.15, 0, 0, 0,), \\
&\quad (1, 1, 0.98, 0.89, 0.78, 0.66, 0.5, 0)); \\
FL_f(\widetilde{A}) &= ((0, 0.02, 0.11, 0.22, 0.34, 0.5, 1, 1), \\
&\quad (0.01, 0.22, 0.34, 0.85, 1, 1, 1, 1); \\
FE_f(\widetilde{A}) &= ((0, 0.02, 0.11, 0.22, 0.15, 0, 0, 0), \\
&\quad (0.01, 0.22, 0.34, 0.85, 0.78, 0.66, 0.5, 0)).
\end{aligned}
\tag{4.44}
$$

Their graphical interpretation has been presented in Fig. 4.18.

As one can see from the examples above, correct selection of cardinality pattern and its generating parameters is of key importance for modelling cardinalities of interval-valued fuzzy sets. In the following Section we will present methods of fitting of such functions to concrete decision problem.

4.6 Decision Making Model Based on Cardinalities of IVFSs

4.6.1 Introduction to Group Decision Making (GDM)

The problem of group decision making occurs in a situation when many experts (decision sources) need to collectively make a common decision (from among available options) on the basis of their own opinions. Two most important models of group decision making include methods based on consensus and based on voting. Methods based on consensus try to reconsolidate contradicting opinions in such a way as to avoid "winners" and "losers". By contrast, in voting-based methods the majority of votes is decisive. The GDM problematic is very popular and rich. Naturally, soft methods work well this issue. A very good overview of consensus-based methods can be found in Herrera-Viedma et al. [55]. In contrast to voting-based methods, where generally negotiations are not taken into account, consensus-based methods assume cooperation and working out common opinions. The approach introduced by Kacprzyk (see Kacprzyk [65–67]) based on applications of Zadeh's linguistic quantifiers (see Zadeh [158]) of the "majority", "almost everybody" type and introducing the notion of "soft consensus" proves especially interesting. It was one of the first applications of methods of counting elements of fuzzy sets to group decision making.

In a classical formulation of a GDM problem (Chen and Hwang [16], Fodor and Roubens [43], Herrera-Viedma et al. [55]) there is a set of possible alternatives $X = \{x_1, \ldots, x_n\}$ and a group of experts $E = \{e_1, \ldots, e_m\}$, who express their opinions on set X wishing to make a common decision. It is possible to differentiate weight of opinions of specific experts by introducing weights describing their relevance in decision process. Decision process can be divided into two stages: selection and consensus reaching. At the selection stage (see Fodor and Roubens [43]) a set of alternatives preferred by experts can be isolated. Usually it is done by aggregating preferences given by experts on their opinion decisions. This might, however, lead to solutions unacceptable by some of the experts and thus consensus reaching stage is often used during which experts discuss and negotiate their decisions.

Further in this work we will focus on voting methods that do not provide for the consensus reaching stage. This is due to the character of our problem in which experts are decision sources in the form of mathematical models.

4.6.2 Decision Algorithms

One of the decision making methods natural for humans in a situation when they have knowledge from various sources is the counting strategy. Depending on situation, data sources may include experts, Internet sources, mass media, academic papers etc. By counting people decide how many sources are for, and how many are against. Usually, in such a situation they assume that the majority is right (democratic approach). However, the approach to understanding majority may differ (how many more votes need to be casted for to make a decision). This decision making method is called voting.

Because, in our case, source decisions are represented by intervals (interval-valued fuzzy set), it seems natural to estimate minimum and maximum conviction for a given decision. Such an approach suggest use of interval-valued fuzzy sets cardinalities to represent boundary of both decision intervals (for decision support and rejection). We must be conscious of the fact that, as noted by Dubois et al. [32], voting approach is justifiable when sources are independent. In our research independent sources should be understood as calculation models coming from various research centres prepared by experts with varied experience. We may thus assume that they are independent and are equally important for the decision.

As described above, the decision model in question assumes data sources that might be uncertain. The decisions are represented as IVFS $\tilde{D} = (D_l, D_u)$ with $D_l \subset D_u$ defined on a finite universe n of decision sources $\{S_1, S_2, \ldots, S_n\}$.

Fuzzy sets D_l and D_u represent lower and upper limit of certainty level for a decision made available by information source. These values may be interpreted in the following manner: values closer to 1 represent an inclination towards a positive decision, whereas values closer to 0 represent an inclination towards negative decision.

Figure 4.19 presents an example of six decision information sources for a given problem.

The idea behind this algorithm conforms with a usual method of making decisions by counting crisp sets. We make a decision supported by majority of data sources, on condition that they are more numerous that the reverse option by a specified value. If both options has the same support of decision sources (or the difference is minimal), then we do not make a decision.

The idea behind decision algorithm is to use bipolar perspective on IVFS. Because such an IVFS contains information on uncertainty level, it carries both information supporting and rejecting the decision. This property of IVFS is used in decision algorithm.

The basic idea behind this algorithm consists of a couple of steps:

- On the basis of input data we define two IVFSs: IVFS "Pro" modelling support level for a positive decision $\tilde{P} = (D_l, D_u)$ and IVFS "Contra" mirroring support level for a negative decision $\tilde{C} = (D'_u, D'_l)$.
- We calculate cardinalities of these IVFSs with the selected calculation method. Version of the algorithm using cardinality *sigma f-Count* has been presented in Sect. 4.6.3. By contrast, the version using cardinality of type *f-FEcount* has been described in Sect. 4.6.4.
- We compare cardinalities to find out whether we can make a decision i.e. whether one of them significantly outweighs the other, and if so we select the decision supported by greater cardinality.

Fig. 4.19 Example of six information sources as IVFS $\tilde{D}$

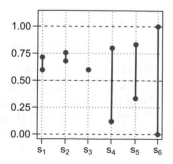

When defining algorithm and comparing intervals it is useful to introduce a notion of interval representative $\tilde{x}$ designated by $Rep(\tilde{x})$. It is a single real number belonging to this interval. The most obvious interval representatives include: interval centre Rep_{cen}, right limit (minimum value) Rep_{min}, left limit (maximum value) Rep_{max}.

Depending on selection of calculation methods and comparison methods we obtain various decision algorithms.

4.6.3   Algorithm Using Scalar Cardinality Sigma F-Count

As described above, first we defined two IVFSs: IVFS "Pro" modelling support level for a positive decision $\tilde{P} = (D_l, D_u)$ and IVFS "Con" mirroring support level for a negative decision $\tilde{C} = (D'_u, D'_l)$.

Then, we calculate cardinalities of both IVFSs using *sigma f-Count* sc_f (see Sect. 4.4) with appropriately selected cardinality pattern f. As a result we obtain two intervals: $sc_f(\tilde{P}) = [\sigma_f(D_l), \sigma_f(D_u)]$ representing the number of sources voting for and interval $sc_f(\tilde{C}) = [\sigma_f((D_u)'), \sigma_f((D_l)')]$ representing the number of sources voting against.

Please note that if we use cardinality patterns $f_{1,0.5,1}$ (*counting by thresholding*) this approach would be equivalent to classical voting strategy (each decision source will return 0 or 1).

In order to make a decision we need to determine a method for comparing cardinality intervals. For this purpose we defined two approaches (modes):

- interval approach consisting in comparing overlap of intervals of respective distances between their endpoints,
- numerical approach consisting of determining numerical interval representatives.

Representatives of cardinality intervals for interval-valued fuzzy set $\tilde{A} = (A_l, A_u)$ are defined in the following manner:

$$
\begin{aligned}
Rep_{min}(\tilde{A}) &= sc_f(A_l), \\
Rep_{max}(\tilde{A}) &= sc_f(A_u), \\
Rep_{cen}(\tilde{A}) &= (sc_f(A_l) + sc_f(A_u))/2.
\end{aligned}
\tag{4.45}
$$

When operating on intervals we can make final decision using (Algorithm 1).

Specific algorithm steps may be described in the following manner:

- If cardinality intervals overlap or the distance between respective interval endpoints is lower than a given value r (i.e. difference in the number of votes between contrasting decisions is insufficient), the algorithm is unable of making decision. The parameter r expresses how much needs to be the difference between sources supporting and rejecting the decision in order to make a final decision.

Algorithm 1 Decision algorithm using interval mode

if $Rep_{max}(\widetilde{P}) \lesseqgtr Rep_{min}(\widetilde{C})$ **then**
 if $Rep_{min}(\widetilde{C}) - Rep_{max}(\widetilde{P}) \geq r$ **then**
  $Decision \leftarrow Negative$
 else
  $Decision \leftarrow NA$
 end if
else
 if $Rep_{max}(\widetilde{C}) \lesseqgtr Rep_{min}(\widetilde{P})$ **then**
  if $Rep_{min}(\widetilde{P}) - Rep_{max}(\widetilde{C}) \geq r$ **then**
   $Decision \leftarrow Positive$
  else
   $Decision \leftarrow NA$
  end if
 else
  $Decision \leftarrow NA$
 end if
end if

- in the other case we select as decision the interval with larger cardinality values. This means that we are selecting decision supported by a sufficient number of decision sources.

In numerical mode the algorithm (only operating on representatives of a single type) gets significantly simpler calculation-wise. It has been presented in the Algorithm 2. It is done for a set representative *Rep* out of representatives defined in (4.45).

Algorithm 2 Decision algorithm for interval-valued scalar cardinality *sigma f-Count* sc_f with the use of numerical mode with given representative *Rep* (4.45).

if $|Rep(\widetilde{P}) - Rep(\widetilde{C})| < r$ **then**
 $Decision \leftarrow NA$
else
 if $Rep(\widetilde{P}) > Rep(\widetilde{C})$ **then**
  $Decision \leftarrow Positive$
 else
  $Decision \leftarrow Negative$
 end if
end if

We will present operation of this algorithms on the example below:

Example 4.7 Let us consider the following IVFS $\widetilde{D} = (D_l, D_u)$ as input for decision algorithm:

$$D_l = 0.55/x_1 + 0.59/x_2 + 0.38/x_3 + 0.12/x_4 + 0.83/x_5 + 0/x_6,$$
$$D_u = 0.55/x_1 + 0.59/x_2 + 0.50/x_3 + 0.78/x_4 + 1/x_5 + 0.65/x_6. \tag{4.46}$$

Fig. 4.20 Graphical
interpretation of IVFS $\tilde{A}$
(4.46)

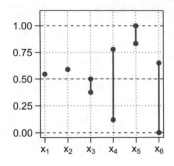

Its graphical representation has been presented in Fig. 4.20
We generate two IVFSs $\tilde{P}$ and $\tilde{C}$

$$P_l = 0.55/x_1 + 0.59/x_2 + 0.38/x_3 + 0.12/x_4 + 0.83/x_5 + 0/x_6,$$
$$P_u = 0.55/x_1 + 0.59/x_2 + 0.50/x_3 + 0.78/x_4 + 1/x_5 + 0.65/x_6. \qquad (4.47)$$

$$C_l = 0.45/x_1 + 0.41/x_2 + 0.50/x_3 + 0.22/x_4 + 0/x_5 + 0.35/x_6,$$
$$C_u = 0.45/x_1 + 0.41/x_2 + 0.62/x_3 + 0.88/x_4 + 0.17/x_5 + 1/x_6. \qquad (4.48)$$

In the examples below let us assume that decision should be made with at least 1
vote lead (parameter $r = 1$).

1. For cardinality pattern $f = f_{id}$ (standard sigma count) we obtain the following
   set representatives $\tilde{P}, \tilde{C}$:

	$\tilde{P}$	$\tilde{C}$
Rep_{min}	2.47	1.93
Rep_{cen}	3.27	2.73

On the basis of Algorithms 1 and 2 we obtain the following decisions:

- No decision (NA)—interval mode (Algorithm 1);
- No decision (NA)—numerical mode (Algorithm 2) for $Rep = Rep_{max}$;
- No decision (NA)—numerical mode (Algorithm 2) for $Rep = Rep_{cen}$.

As one can see neither of the variants was capable of making a decision.

2. For cardinality pattern $f = f_{3,0.5,07,1}$ we obtain the following IVFS representatives $\tilde{P}, \tilde{C}$:

	$\tilde{P}$	$\tilde{C}$
Rep_{min}	2.14	0
Rep_{max}	3.79	2.62
Rep_{cen}	2.96	1.31

On the basis of Algorithms 1 and 2 we obtain the following decisions:

- No decision (NA)—interval mode (Algorithm 1);
- Positive decision(1)—numerical mode (Algorithm 2) for $Rep = Rep_{max}$;
- Positive decision(1)—numerical mode (Algorithm 2) for $Rep = Rep_{cen}$.

As you can see, interval mode is much more restrictive and only efficient in situations with small amount of missing information (size of ignorance intervals). Definition of cardinality pattern is also of key importance. If using identity function as cardinality pattern (using *sigmacount* for calculating cardinality) cardinalities of both IVFSs $\widetilde{P}$ i $\widetilde{C}$ are symmetrical and decision is only made if both IVFSs are sufficiently similar to crisp sets. It is also important that thresholds (parameters defining cardinality pattern) be selected in such a way as to reduce the significance of input decisions close to 0.5.

4.6.4 Algorithm Using Scalar Cardinality f-FEcount

As in the previous version of the algorithm we start by defining two IVFSs: IVFS "Pro" modelling support level for a positive decision $\widetilde{P} = (D_l, D_u)$ and IVFS "Contra" mirroring support level for a negative decision $\widetilde{C} = (D'_u, D'_l)$.

Then, we calculate fuzzy cardinalities *f-FECount* FE_f of both IVFSs (see 4.28) with appropriately selected cardinality pattern f. As a result we obtain two fuzzy cardinalities: $FE_f(\widetilde{P}) = (FE_f(P_l), FE_f(P_u))$ representing the number of sources voting "for" $FE_f(\widetilde{C}) = (FE_f(C_l), FE_f(C_u))$ representing the number of sources voting "against". For establishing cardinality intervals (defuzzification) we use function COG (3.38).

Representatives are defined differently than in the case of scalar cardinalities. In this case we define them in the following manner:

$$
\begin{aligned}
Rep_{min}(\widetilde{A}) &= min(COG(FE_f(A_l), COG(FE_f(A_u)))), \\
Rep_{max}(\widetilde{A}) &= max(COG(FE_f(A_l), COG(FE_f(A_u)))), \quad (4.49) \\
Rep_{cen}(\widetilde{A}) &= [COG(FE_f(A_l)) + COG(FE_f(A_u))]/2.
\end{aligned}
$$

Then, we make a decision using decision algorithms defined in the previous section: Algorithm 2—in interval mode and Algorithm 2—in numerical mode with a single set representative Rep (4.49). Principle of operation of the decision algorithms is identical to the rule described in Sect. 4.6.3.

Let us consider an example of algorithm operating for an input set $\widetilde{D}$ (4.46) from Example 4.7 and various decision parameters i.e. cardinality patterns, decision methods and various representatives.

Example 4.8 In the examples below let us assume that decision should be made with at least one vote lead (parameter $r = 1$).

1. For cardinality pattern $f = f_{id}$ we obtain the following representatives of cardinalities of IVFS $\widetilde{P}$, $\widetilde{C}$:

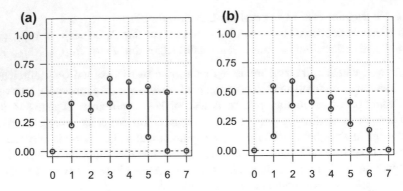

Fig. 4.21 Cardinalities of IVFSs **a** $FE_f(\widetilde{P})$, **b** $FE_f(\widetilde{C})$

	$\widetilde{P}$	$\widetilde{C}$
Rep_{min}	2.89	3.03
Rep_{max}	3.62	3.11
Rep_{cen}	3.25	3.07

Cardinalities FE_f of IVFSs $\widetilde{P}$ and $\widetilde{C}$ have been presented in graphs in Fig. 4.21. On the basis of Algorithms 1 and 2 we obtain the following decisions:

- No decision (NA)—interval mode (Algorithm 1);
- No decision (NA)—numerical mode (Algorithm 2) for $Rep = Rep_{max}$;
- No decision (NA)—numerical mode (Algorithm 2) for $Rep = Rep_{cen}$.

As one can see neither of the variants was capable of making a decision.

2. For cardinality pattern $f = f_{3,0.5,07,1}$ we obtain the following set representatives $\widetilde{P}, \widetilde{C}$:

	$\widetilde{P}$	$\widetilde{C}$
Rep_{min}	2.54	0
Rep_{max}	3.16	1.85
Rep_{cen}	2.85	0.93

Cardinalities FE_f of IVFSs $\widetilde{P}$ and $\widetilde{C}$ have been presented in graphs in Fig. 4.22. On the basis of Algorithms 1 and 2 we obtain the following decisions:

- No decision (NA)—interval mode (Algorithm 1);
- Positive decision (1)—numerical mode (Algorithm 2) for $Rep = Rep_{max}$;
- Positive decision (1)—numerical mode (Algorithm 2) for $Rep = Rep_{cen}$.

Similarly as in the previous example, we can see that the interval mode is more restrictive in decision making. Intervals generated by defuzzyfication function COG for fuzzy cardinalities of type f-$FECount$ are narrower and they provide a better

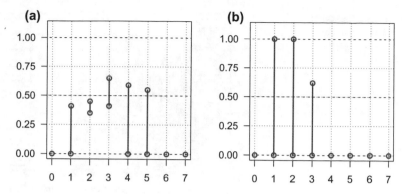

Fig. 4.22 Cardinalities of IVFSs **a** $FE_f(\widetilde{P})$, **b** $FE_f(\widetilde{C})$

approximation of cardinalities of decision sets even with relatively large level of ignorance in input IVFSs.

Thanks to application of fuzzy approach to calculating cardinalities of corresponding interval-valued fuzzy sets we obtain a better representation of cardinality and thus a better potential decision efficacy.

4.7 Evaluation of Algorithms Efficacy

The presented algorithms have been tested on real medical data. These data described 388 cases of patients diagnosed and treated in the Division of Gynecological Surgery, Poznan University of Medical Sciences, between 2005 and 2015. Out of them 61% have been diagnosed as suffering from benign tumours and 39% as suffering from malign tumours. Moreover, 56% of patients had full diagnostic (no test required by diagnostic scales was missing), 40% had significant amounts of missing data varying from (0%, 50%], and for the remaining ones 50% of data was missing. Percentage breakdown of missing values has been presented in Fig. 4.23. Detailed description of data used for evaluation can be found in Moszyński et al. [86]. More information on the data format used and technical details can be found in Wójtowicz et al. [151].

The following six diagnostic models have been used for evaluation: SM (Sonomorphological index) (see Sect. 2.2.4.9), Alc (Alcazar's model) (see Sect. 2.2.4.2), IOTA LR1 and LR2 (see Sect. 2.2.4.4), Tim. (Timmerman's logistic regression model) (see Sect. 2.2.4.10), RMI (Risk of malignancy index) (see Sect. 2.2.4.6). Table 4.1 presents models and diagnostic attributes used by them. Diagnostic attributes have been divided into two groups: the first one resulting from medical interview is always complete, whereas the second one may include missing data. The methods have been subjected to uncertaintification described in Sect. 4.7.2.

Fig. 4.23 Distribution of
patients with respect to
percentage-wise gaps in
diagnostic data

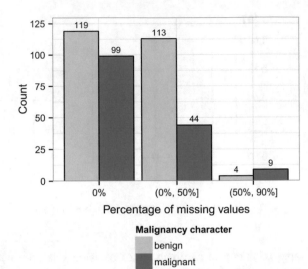

Table 4.1 Diagnostic attributes used for selected diagnostic models

Attribute	Diagnostic models					
	SM	Alc.	LR1	LR2	Tim.	RMI
	m_1	m_2	m_3	m_4	m_5	m_6
Age	-	-	✓	✓	-	✓
Menopausal status	✓	-	-	-	✓	✓
Pain during examination	-	-	✓	-	-	-
Hormonal therapy	-	-	✓	-	-	-
Hysterectomy	-	-	-	-	-	✓
Ovarian cancer in family	-	-	✓	-	-	-
Lesion volume	✓	-	✓	-	-	-
Internal cyst walls	✓	-	✓	✓	-	-
Septum thickness	✓	-	-	-	-	-
Echogenicity	✓	✓	-	-	-	-
Localisation	✓	-	-	-	-	✓
Ascites	✓	-	✓	✓	-	✓
Papillary projections	-	✓	-	-	✓	-
Solid element size	-	✓	✓	✓	-	✓
Blood flow location	-	✓	✓	✓	-	-
Resistance index	-	✓	-	-	-	-
Acoustic shadow	-	-	✓	✓	-	-
Amount of blood flow	-	-	✓	-	✓	-
CA-125 blood marker	-	-	-	-	✓	✓
Lesion quality class	-	-	-	-	-	✓

4.7.1 Decision Intervals Modelling

In the case of classical decision systems, patient is modelled as vector p in universe P. Set P is defined as $P = D_1 \times \ldots \times D_n$, where $D_1, \ldots, D_n$ are closed intervals of real numbers designating domains of attributes describing patients. The vector p describing a patient takes a form $p = (p_1, \ldots, p_n)$, where $p_i \in D_i$.

Diagnostic model may be formalized as function $m : P \rightarrow [0, 1]$. Values returned by the function pertain to certainty degree as to malign character of the tumour as is interpreted in the following manner:

1. $m(p) \geq 0.5$—decision towards malign character of the tumour (larger values indicate more certainty);
2. $m(p) < 0.5$—decision towards benign character of the tumour (lower values indicate more certainty);

Please note that a situation when $m(p) = 0.5$ is interpreted as decision towards malign character of the tumour.

There are diagnostic models operating on complete patient data and unable to make any decisions in case of partial gaps. In order to be able to represent such gaps, an additional symbol has been introduced (in practice denoted as NA) into the domain of each of the attributes. Thus each of the patients is described by vector $p = (p_1, \ldots, p_n)$, where $p_i \in D_i \cup \{NA\}$. The disadvantage of this solution is the fact that original diagnostic models are unable of dealing with such values. Therefore, we decided to use another approach in which all the data will be represented in the same, coherent manner (see Żywica et al. [166]).

Each attribute D_i is substituted with a corresponding interval defined as a set of all non-empty closed sub-intervals in D_i

$$\widetilde{D_i} = \mathcal{I}_{D_i} = \{[a, b] : [a, b] \subseteq D_i\} . \tag{4.50}$$

Similarly as above, we define $\widetilde{P} = \tilde{D}_1 \times \tilde{D}_2 \times \ldots \times \tilde{D}_n$. As in the remaining parts of the book, we use the tilde character to designate interval value as opposed to numerical value.

In such a model a patient is described as a vector of intervals

$$\widetilde{p} = (\widetilde{p}_1, \ldots, \widetilde{p}_n) = \left([\underline{p}_1, \overline{p}_1], \ldots, [\underline{p}_n, \overline{p}_n]\right) . \tag{4.51}$$

We say that vector $p \in P$ is an embedded vector of the vector $\widetilde{p} \in \widetilde{P}$ designated with $p \in_E \widetilde{p}$ and we write

$$\forall_{1 \leq i \leq n} \ p_i \in \widetilde{p}_i . \tag{4.52}$$

For each vector $p \in P$ (both with and without missing data) we may define its interval equivalent $\widetilde{p} \in \widetilde{P}$ in the following manner:

$$\underline{p}_i = \begin{cases} p_i & \text{if } p_i \neq NA \\ \min\limits_{d \in D_i} d & \text{if } p_i = NA \end{cases}, \quad \overline{p}_i = \begin{cases} p_i & \text{if } p_i \neq NA \\ \max\limits_{d \in D_i} d & \text{if } p_i = NA. \end{cases} \tag{4.53}$$

The definition above allows us to describe attribute values with an interval, regardless of whether the attribute value has been given. If attribute value has not been given, then in the proposed representation it is substituted with an interval including all possible attribute values. If the value has been provided, it will be represented with a single numerical value. The major advantage of this approach is that data of all patients are represented in the same, uniform manner and processed by all existing diagnostic models.

4.7.2 Uncertaintification of Diagnostic Models

The standard diagnostic models operate only on numerical values. In order to allow them to make decisions based on new interval representation, introduced in Sect. 4.7.1, they need to be uncertaintificated. For this purpose we will use a classical method of expanding a real function into an interval (see Moore [82]). The new, uncertaintified diagnostic model $\widetilde{m}$ is defined as:

$$\widetilde{m}(\widetilde{p}) = \{m(p) : p \in_E \widetilde{p}\} . \tag{4.54}$$

The resulting interval represents all possible diagnoses that may be made based on patient descriptions in which each missing value has been substituted with all possible values of this attribute. The more the description is imperfect, the more imprecise the diagnosis. Please note, that in many cases it will be still possible to make a correct decision as a certain amount of missing values is allowable and does not influence the final decision in a significant manner.

It is also natural to expect that reasoning based on an interval will return an interval. Therefore, the value $\widetilde{m}(\widetilde{p})$ can also take the form of an interval:

$$\widetilde{m}(\widetilde{p}) = \left[\min_{p \in_E \widetilde{p}} m(p), \max_{p \in_E \widetilde{p}} m(p) \right] . \tag{4.55}$$

Both definitions are equivalent if original diagnostic models are continuous (e.g. in the case of models based on linear or logistic regression). If the model m is not continuous, then (4.55) gives a very good approximation of (4.54), and therefore we accept the formula (4.55) as a definition of the new (uncertaintified) model $\widetilde{m} : \widetilde{P} \rightarrow \mathcal{I}_{[0,1]}$.

4.7.3 Decision Making Procedure

We assume that in decision making process we will use efficient and well-known diagnostic models. Using interval representation and the process of uncertaintifi-cation we are able to adapt these models to operation with imperfect data. Thus, using definitions from previous sections for each patient we are able to construct an interval-valued fuzzy set on a universe of allowable diagnostic models and thus use decision methods operating on interval-valued fuzzy sets. Thanks to such an ap-proach we are able to increase efficacy of decision making and reduce the influence of missing data on final decision. Additionally, we accept a situation in which we will not be able to make a final decision which prevents us from making a wrong decision in the case of lack of sufficient amount of data.

The proposed approach for making medical decision has been presented in Fig. 4.24.

In the OvaExpert system we also implemented other decision methods based of data aggregation (see Sect. 5.2.1) and interval classifier (see Sect. 5.2.2). Comparison

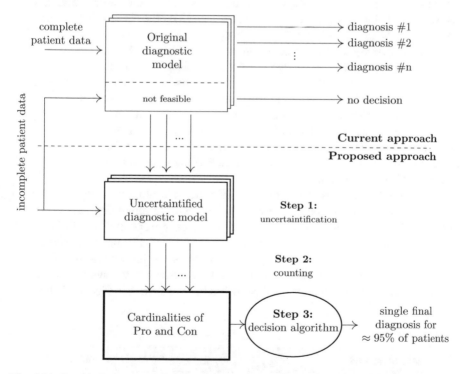

Fig. 4.24 Graphical summary of the current and the proposed approach. *Rectangles* represent diagnostic models on various stages. *Vertical arrows* designate transformations of diagnostic models (uncertaintification and counting). The third stage (decision) has been represented as an ellipsis. *Horizontal arrows* designate flow of patient data and diagnosis

of their decision efficacy on test data and conclusions from their applications have
been presented in Sect. 5.2.4.

4.7.4 Evaluation

Model evaluation procedure was based on classical division of data into two sets:
training and test ones. The original set of data included few patients with lacking
attributes at each of levels of lack of data. Therefore, we could not use it directly
for training and testing our methodologies as it would lead to distortions in data
distribution. We could try to increase the set by including new cases, but this would
yield a significant delay in research and necessity of collecting new data. Thus, we
decided to use another approach.

We established that test set would include data of patients with real gaps and a
certain portion of patients with complete data. The training set was construed on the
basis of set with complete descriptions in which we simulated data deletion. In our
incompleteness simulation we assumed that gaps will be at a certain random level
as we are not able to mirror real level of gaps caused by the diagnostic process. Real
distribution of gaps in attributes describing patients is unknown, thus we decided to
simulate uniformly all the gap levels in the training phase. As a consequence both
sets, training and testing ones, do not have discontinuities in data gap distributions.

Additionally, real distribution of malign tumours in the population remains unknown. Some statistics relating to this issue may be found in Stukan et al. [117],
where the author present various prediction models with tumour malignity distributions in specific source research. Malignity/benignity coefficients are very varied in
specific groups. Thus, in the simulation process we assumed that malignity distribution is constant i.e. we are randomly selecting the same number of patients with
malign and benign tumours.

The training set consisted of data of 200 patients without missing attributes required by diagnostic models. The test set included the remaining data of 18 patients without missing data and such for which gaps were included in the interval (0%, 50%]. In such a way we construed a test set of 175 patients. The above-
mentioned subgroups of 200 and 18 cases had the same proportion of malign and
benign tumours. Patients with more than 50% of missing data have been excluded
from our research. Distributions of data set has been presented in Figs. 4.25 and 4.26.

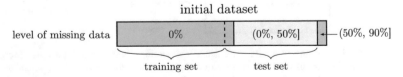

Fig. 4.25 Division of data sets. Data of patients with more than 50% of missing data have not been
taken into account in the experiment

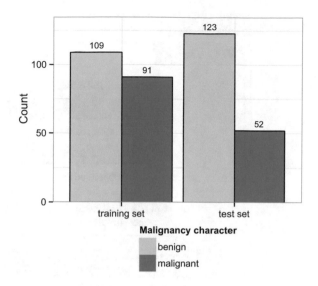

Fig. 4.26 Distribution of malignity classes in training and test sets

The goal of the training phase was to fit decision algorithms described in Sect. 4.6.2 i.e. selection of a decision method and selection of cardinality patterns and parameters to achieve the best possible decisions parameters (minimizing the cost functions) for various levels of missing data. Missing data level was set at various levels varying from 0 to 50% with 5% step. For each missing data level we performed 1000 repetitions of the following procedure. In the first step, we randomly selected 75 patients with malign tumour and 75 patients with benign tumour from the training set and we deleted a given percentage of attribute values randomly from their data. Then, for such data we calculated uncertaintified diagnostic models returning interval-valued diagnoses. Finally, we run decision algorithm calculating the final decision for specific parameters. For decisions obtained in such a way we calculated quality measures described in Sect. 2.3.2 and averaged all the results from all the missing data levels. In order to avoid over-learning effect, the algorithms was run on a reasonable set of numerical parameters selected by experts. Specific steps of the training phase have been presented in Fig. 4.27.

As a result of the training phase we obtain a set of optimized decision algorithms that achieved very good results on training data (with simulation with a certain amount of missing data).

In the testing phase the optimized decision algorithms have been tested with regard to decision quality on a test data. In this step decision quality was tested on data including real gaps. In order to establish confidence intervals of the quality measures we used stratified bootstrapping with 500 repetitions. More on this method can be found in Japkowicz and Shah [64], Davison and Hinkley [24].

The goal of evaluation was to select a decision algorithm that would best classify malignity cases with the top possible decisiveness. As it has been presented in Sect. 2.3 there are many measures of classification quality. In the medical problem

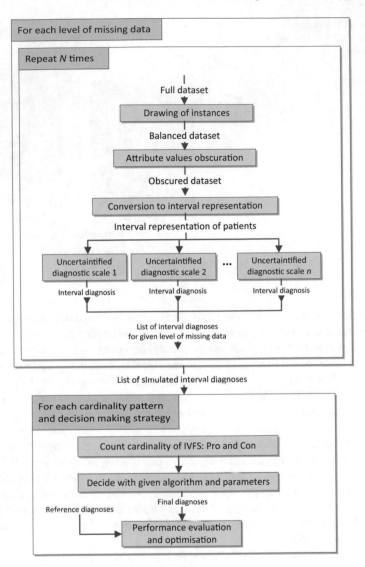

Fig. 4.27 Visualization of the training phase. Data flow is represented by *arrows* and *rectangles* represent operations on data

under consideration high sensitivity and specificity is crucial. Moreover, in a certain number of cases diagnostic models are not unambiguous and thus, the decision method should not run classification by chance. In such a case, decision should not be taken as the patient should be subjected to additional diagnostic tests or referred to a more experienced specialist. Therefore, we accept a situation that some cases remain unclassified (decisivness below 100%), see Sect. 2.3.2.7). Additionally, selection of

Table 4.2 Cost matrix used for models assessment

Actual	Predicted		
	Benign	Malignant	NA
Benign	0	2.5	1
Malignant	5	0	2

a right, unified quality measure constitutes a difficult task (see Japkowicz and Shah [64]). Thus, we decided to use a cost matrix for assessment (see Sect. 2.3.2.8).

Specific value of costs matrix have been selected in cooperation with experts in ovarian cancer diagnosis. The Table 4.2 presents costs (penalties) attributed to classifiers for incorrect decisions. Correct decisions (TP and TN) do not receive a penalty. A classifier receives top penalty in the case of committing type II error (FN) i.e. if a patient with malign tumour is classified as a benign case. Penalty for FP type errors was half of it, as unjustified operation is still dangerous for a patient but death risk is much lower. Additionally, there are also penalties for the classifier for failure to make a decision (NA). The penalty is lower, as in such a case the patient needs additional diagnostics and will probably be directed to a more experienced specialist who would make a correct diagnosis. However, we differentiated penalties for lack of decision in positive (malign) is twice as high as in the negative (benign) case.

Statistical evaluation and implementation of the proposed methods have been performed with R, version 3.1.2 (R Core Team [108]). Scripts, documentation and non-sensitive data are available at GitHub (see OvaExpert Project [94]). Because of

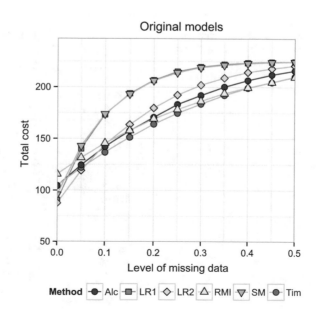

Fig. 4.28 Results achieved by original diagnostic models (with respect to total costs) with various missing data levels

large amount of calculations needed to do the research, we did them using Micrsoft
Azure cloud service (see Microsoft Corp. [81]) available to our team under Microsoft
Azure Research Grant "Azure Machine Learning—Development of an Intelligent
System for Ovarian Tumour Diagnosis".

4.7.5 Results

At the training stage we performed calculations for both methods of calculating
cardinalities of IVFSs (scalar and fuzzy) with the use of five different weighting
functions (cardinality patterns) described in Sect. 3.4.1 with which we tested three
decision algorithms: one interval-valued and two based on representatives. Rep_{cen}
and Rep_{max} with various parameters r determining advantage of votes. In Tables 4.3
and 4.4 we presented results respectively for cardinality of type *sigma f-Count* and
f-FECount broken down into 3 groups of decision algorithms. In each group of
algorithms the best decision result with respect to cost function (lower total costs)
for each of the five cardinality patterns have been presented.

Figures 4.28, 4.29, 4.30 and 4.31 present a summary of the experiment done on
training data with respect to total costs and missing data level.

The Fig. 4.28 presents efficacy of individual original diagnostic models. In this
place we need to explain that they could only make a decision if no data was missing
in a set used by a given model and this decisiveness values are very low (see Fig. 4.32).
Please note that, as expected, their efficacy drops rapidly as the amount of missing
data increases.

Fig. 4.29 Results achieved
by uncertaintified diagnostic
models (with respect to total
costs) with various missing
data levels

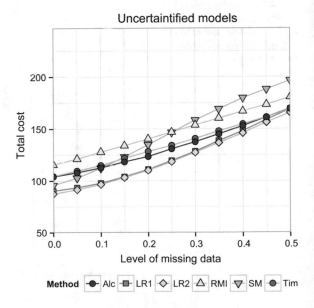

The Fig. 4.29 presents decision efficacy of uncertaintified models. In this case the costs increases (quality drops) much slower. This shows that the mere use of uncertaintification allows for reducing the influence of missing data.

Figures 4.30 and 4.31 present classification results based on the proposed algorithms with the best versions obtained from optimization in specific groups. The graph in Fig. 4.30 presents the best three algorithms based on *sigma f-Count* (see 4.6.3) from specific groups: SC-cen—based on Rep_{cen}, SC-int—based on Rep_{int}, SC-max—based on Rep_{max}. Whereas, the graph presented in Fig. 4.31 presents three best algorithms based on *f-FEcount* (see 4.6.4) from specific groups: FE-cen—based on Rep_{cen}, FE-int—based on Rep_{int}, FE-max—based on Rep_{max}. One can see that diagnostic efficacy of both algorithms is much better in the case of individual models for all levels of missing data. One can also see that some of the algorithms are more sensitive to missing data e.g. *FE-max*.

Algorithms selected in the training phase were further run on test data i.e. real data with real (not simulated gaps). Figures 4.32 and 4.33 have been presented to compare results of classification efficacy with these data for original and uncertaintified models respectively.

Figure 4.34 presents detailed results taking into account various quality measures for counting-based algorithms. These results confirm high efficacy of the proposed algorithms in classifying this type of medical data with gaps. As one can see, original models achieve relatively high values for quality measures such as precision, specificity and sensitivity. However, this is done with very low decisiveness values which means that they only make decisions in few cases. At the same time, the proposed counting algorithms achieve very high value of these parameters with very high decisiveness.

Fig. 4.30 Results achieved by algorithms based on cardinality *sigma f-Count*. The graphs presents one algorithms from each group

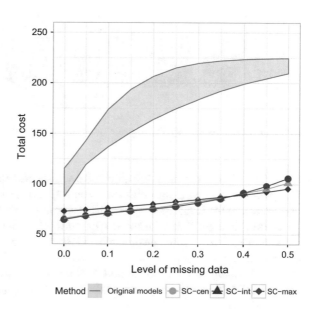

Fig. 4.31 Results achieved
by algorithms based on
f-FEcount. The graphs
presents one algorithms from
each group

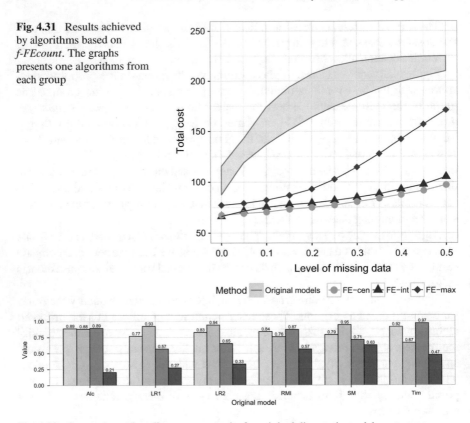

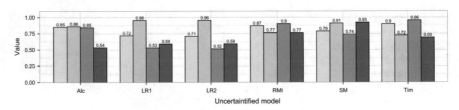

Fig. 4.32 Comparison of quality measure results for original diagnostic models on test set

Fig. 4.33 Comparison of quality measure results for uncertaintified diagnostic models on test set

As a result of analysis of the obtained decision efficacy, the algorithm *FE-cen* (*f-FECount* together with algorithm based on Rep_{cen}) has been selected as the best for application in the OvaExpert system from amount the counting methods. A method based on this algorithm with the use of cardinality pattern $f_{3,0.4,0.9,1}$ and decision parameter $r = 0.5$ will be designated as *FSC* (Ovaexpert system method based on counting).

Results obtained with the method *FSC* are a bit better than those obtained with method *OEA* (OvaExpert system method based on aggregation operators). This is

Table 4.3 Decision efficacy results for algorithms based on scalar cardinality of type *sigma f-Count* for various decision algorithms and cardinality patters with bootstrap percentile 95% confidence intervals, achieved on test data. Decisiveness, sensitivity, specificity are expressed in percentages

Card. Pattern f	r	Performance maeasures with 95% CI				
		Total cost	Acc	Dec	Sen	Spec
f_{id}	0	98.5($\pm$22.6)	93.5($\pm$4.2)	70.9($\pm$6.0)	90.6($\pm$10.7)	94.6($\pm$4.7)
$f_{1,0.5,1}$	0	80.0($\pm$23.0)	89.0($\pm$4.9)	82.9($\pm$5.3)	95.7($\pm$5.3)	85.7($\pm$6.8)
$f_{2,0.4,0.25}$	0	97.5($\pm$23.2)	91.4($\pm$4.9)	73.1($\pm$6.6)	94.1($\pm$7.4)	90.4($\pm$6.4)
$f_{3,0.55,0.6,0.5}$	0	99.0($\pm$24.5)	89.3($\pm$5.1)	74.9($\pm$6.0)	94.6($\pm$6.9)	87.2($\pm$6.9)
$f_{4,0.4,0.5}$	0	96.0($\pm$24.2)	89.5($\pm$5.0)	76.0($\pm$6.0)	94.7($\pm$6.7)	87.4($\pm$6.9)
$f_{5,0.5,40}$	0	100.0($\pm$23.7)	90.6($\pm$4.9)	72.6($\pm$6.3)	94.3($\pm$7.3)	89.1($\pm$7.0)
Rep_{cen}						
f_{id}	0.25	80.5($\pm$30.4)	89.0($\pm$5.0)	93.1($\pm$3.7)	84.8($\pm$11.3)	90.6($\pm$5.5)
$f_{1,0.4,1}$	0.5	75.5($\pm$29.4)	87.6($\pm$5.0)	96.6($\pm$2.3)	88.0($\pm$9.5)	87.4($\pm$6.2)
$f_{2,0.55,1}$	0.25	74.0($\pm$28.2)	88.5($\pm$5.1)	94.3($\pm$3.2)	89.6($\pm$8.7)	88.0($\pm$6.2)
$f_{3,0.2,0.7}$	0.5	77.0($\pm$26.9)	91.0($\pm$4.5)	89.1($\pm$4.6)	86.4($\pm$10.5)	92.9($\pm$4.9)
$f_{4,0.55,0.7}$	0.5	78.5($\pm$29.1)	86.9($\pm$5.2)	96.0($\pm$2.6)	89.6($\pm$8.7)	85.8($\pm$6.6)
$f_{5,0.5,10}$	0.25	75.5($\pm$28.6)	87.6($\pm$4.7)	96.6($\pm$2.6)	88.0($\pm$9.0)	87.4($\pm$6.2)
Rep_{max}						
f_{id}	0.25	80.5($\pm$30.4)	89.0($\pm$5.0)	93.1($\pm$3.7)	84.8($\pm$11.3)	90.6($\pm$5.5)
$f_{1,0.6,1}$	0.75	87.5($\pm$31.0)	87.1($\pm$5.2)	93.1($\pm$3.4)	83.7($\pm$10.8)	88.6($\pm$5.7)
$f_{2,0.55,1}$	0.25	82.0($\pm$32.2)	87.4($\pm$5.3)	95.4($\pm$3.1)	85.4($\pm$10.7)	88.2($\pm$5.8)
$f_{3,0.2,0.7}$	0.5	80.5($\pm$26.5)	89.8($\pm$4.9)	89.7($\pm$4.2)	88.1($\pm$10.2)	90.4($\pm$5.5)
$f_{4,0.55,0.7}$	0.5	83.0($\pm$31.1)	87.0($\pm$5.4)	96.6($\pm$2.6)	84.0($\pm$10.0)	88.2($\pm$6.2)
$f_{5,0.5,10}$	0.25	75.5($\pm$28.6)	87.6($\pm$4.7)	96.6($\pm$2.6)	88.0($\pm$9.0)	87.4($\pm$6.2)

the basic decision method in OvaExpert, described in Sect. 5.2.1 and in Żywica et al. [166].

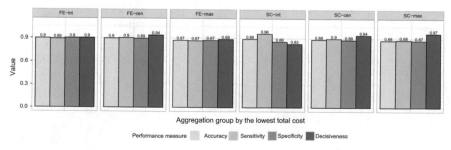

Fig. 4.34 Results of prediction efficecy indicators achieved by individual algorithms on test data

Fig. 4.35 Comparison of
decision power of methods
FSC and OEA on training
data depending on amount of
missing data. Efficacy area
of original models has been
marked in *grey*

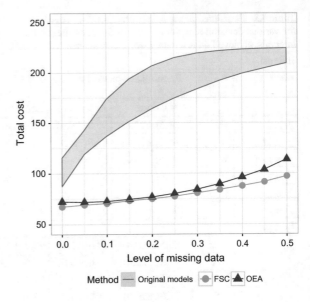

Fig. 4.36 Efficacy of
decision methods *FSC* and
OEA in comparison to
efficacy of original
diagnostic models on test
data

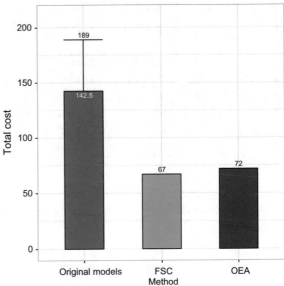

Table 4.4 Decision efficacy results for algorithms based on fuzzy cardinality of type *f-FECount* for various decision algorithms and cardinality patters with bootstrap percentile 95% confidence intervals, achieved on test data. Decisiveness, sensitivity, specificity are expressed in percentages

Card. Pattern f	r	Performance maeasures with 95% CI				
		Total cost	Acc	Dec	Sen	Spec
Interval						
f_{id}	0	98.5($\pm$22.6)	93.5($\pm$4.2)	70.9($\pm$6.0)	90.6($\pm$10.7)	94.6($\pm$4.7)
$f_{1,0.5,1}$	0	80.0($\pm$23.0)	89.0($\pm$4.9)	82.9($\pm$5.3)	95.7($\pm$5.3)	85.7($\pm$6.8)
$f_{2,0.4,0.25}$	0	97.5($\pm$23.2)	91.4($\pm$4.9)	73.1($\pm$6.6)	94.1($\pm$7.4)	90.4($\pm$6.4)
$f_{3,0.55,0.6,0.5}$	0	99.0($\pm$24.5)	89.3($\pm$5.1)	74.9($\pm$6.0)	94.6($\pm$6.9)	87.2($\pm$6.9)
$f_{4,0.4,0.5}$	0	96.0($\pm$24.2)	89.5($\pm$5.0)	76.0($\pm$6.0)	94.7($\pm$6.7)	87.4($\pm$6.9)
$f_{5,0.5,40}$	0	100.0($\pm$23.7)	90.6($\pm$4.9)	72.6($\pm$6.3)	94.3($\pm$7.3)	89.1($\pm$7.0)
Rep_{cen}						
f_{id}	0.25	80.5($\pm$30.4)	89.0($\pm$5.0)	93.1($\pm$3.7)	84.8($\pm$11.3)	90.6($\pm$5.5)
$f_{1,0.4,1}$	0.5	75.5($\pm$29.4)	87.6($\pm$5.0)	96.6($\pm$2.3)	88.0($\pm$9.5)	87.4($\pm$6.2)
$f_{2,0.55,1}$	0.25	74.0($\pm$28.2)	88.5($\pm$5.1)	94.3($\pm$3.2)	89.6($\pm$8.7)	88.0($\pm$6.2)
$f_{3,0.2,0.7}$	0.5	77.0($\pm$26.9)	91.0($\pm$4.5)	89.1($\pm$4.6)	86.4($\pm$10.5)	92.9($\pm$4.9)
$f_{4,0.55,0.7}$	0.5	78.5($\pm$29.1)	86.9($\pm$5.2)	96.0($\pm$2.6)	89.6($\pm$8.7)	85.8($\pm$6.6)
$f_{5,0.5,10}$	0.25	75.5($\pm$28.6)	87.6($\pm$4.7)	96.6($\pm$2.6)	88.0($\pm$9.0)	87.4($\pm$6.2)
Rep_{max}						
f_{id}	0.25	80.5($\pm$30.4)	89.0($\pm$5.0)	93.1($\pm$3.7)	84.8($\pm$11.3)	90.6($\pm$5.5)
$f_{1,0.6,1}$	0.75	87.5($\pm$31.0)	87.1($\pm$5.2)	93.1($\pm$3.4)	83.7($\pm$10.8)	88.6($\pm$5.7)
$f_{2,0.55,1}$	0.25	82.0($\pm$32.2)	87.4($\pm$5.3)	95.4($\pm$3.1)	85.4($\pm$10.7)	88.2($\pm$5.8)
$f_{3,0.2,0.7}$	0.5	80.5($\pm$26.5)	89.8($\pm$4.9)	89.7($\pm$4.2)	88.1($\pm$10.2)	90.4($\pm$5.5)
$f_{4,0.55,0.7}$	0.5	83.0($\pm$31.1)	87.0($\pm$5.4)	96.6($\pm$2.6)	84.0($\pm$10.0)	88.2($\pm$6.2)
$f_{5,0.5,10}$	0.25	75.5($\pm$28.6)	87.6($\pm$4.7)	96.6($\pm$2.6)	88.0($\pm$9.0)	87.4($\pm$6.2)

Graph in Fig. 4.35 presents a comparison in efficacy of algorithms *FSC* and *OEA* on training data depending on the amount of missing data. A slightly better efficacy of algorithm *FSC* can be observed at each missing data level with growing advantage with increase in amount of missing data.

The Fig. 4.36 presents results for test data in comparison to original methods. Total cost generated by algorithm *FSC* is much lower than in original models and remarkably lower than algorithm *OEA*.

Chapter 5
OvaExpert System

My argument is that uncertainty is the great unspoken secret of medicine and that by ignoring this fundamental uncertainty we are doing real harm to ourselves.

Hatch [52]

This chapter details the implementation of the intelligent decision support system OvaExpert for the diagnosis of ovarian tumors. The individual diagnostic modules are discussed and an analysis of the efficacy of the decision-making methods is presented.

5.1 Features of the OvaExpert System

OvaExpert is an intelligent decision support system for the diagnosis of ovarian tumors. It provides an effective diagnosis, factoring in low quality uncertain data, both imprecise and incomplete. The system was developed as a result of joint research of two Polish research centers: the Division of Gynecologic Surgery of the Poznan University of Medical Sciences and the Department of Imprecise Information Processing Methods, Faculty of Mathematics and Computer Science of Adam Mickiewicz University in Poznań.

The main purpose of creating OvaExpert was to equip physicians with an effective tool supporting the following areas:

- reducing the impact of low quality data on the correct diagnosis process,
- presenting diagnostic results in the most intuitive and comprehensive way for the physician,
- collecting and managing data on the diagnosed patients, and storing data in a standardized uniform way.

© Springer International Publishing AG 2018
K. Dyczkowski, *Intelligent Medical Decision Support System Based on Imperfect Information*, Studies in Computational Intelligence 735, https://doi.org/10.1007/978-3-319-67005-8_5

At the moment the OvaExpert system is tested in several medical centers offering the diagnosis and treatment of gynecological tumors. The demo version of the system is available on the project website (see OvaExpert Project Homepage [95]) where one can get acquainted with the functions and possibilities offered by the system.

OvaExpert uses the existing knowledge about ovarian tumors (i.e. existing models, scoring systems, reasoning schemes) and integrates it into a single information system. This system is unique for many reasons. It is the first system that uses imperfect and imprecise data to differentiate ovarian tumors at several levels (see Stachowiak et al. [113]):

- at the stage of collecting patient data,
- at the stage of data processing,
- at the state of diagnosis presentation.

Below is a brief description of the various elements of the system. The system covers four main areas of application: medical data collection, expert knowledge collection, decision support in choosing the optimal diagnostic path, and decision support in determining the type of cancer.

5.1.1 Medical Data Gathering

One of the main goals of the OvaExpert system is to provide a simple, convenient and efficient way to collect data on the diagnosed patients. At present, because of the lack of uniform data collection standards and the lack of a centralized medical gathering system, the cooperation between doctors from different research centers is very limited. If data are collected in such centers, it is usually in their own in-house format, using mostly spreadsheets or paper. Many physicians also do not attach importance to the uniformity of description and its quality (often due to lack of time for note-taking during patient examination). So when there is an attempt to combine these data, ambiguity arises and gaps are found in patient descriptions. To solve these problems, OvaExpert provides a standardized data storage format that complies with the IOTA group guidelines (see Timmerman et al. [133]). With the SaaS (Software as a Service) model, a doctor has access to a central database from any device, such as a PC, tablet or smartphone. Centralizing data facilitates knowledge gathering from different specialised medical centers at the same time, and thus provides data for future research and improvement of decision-making support algorithms.

5.1.2 Easy and Intuitive User Interface

The user interface was developed in collaboration with gynecologists to make sure it was usable for them and also to make it possible to conveniently enter data and access the diagnosis both in the doctor's office (on PC, via the web interface) and outside

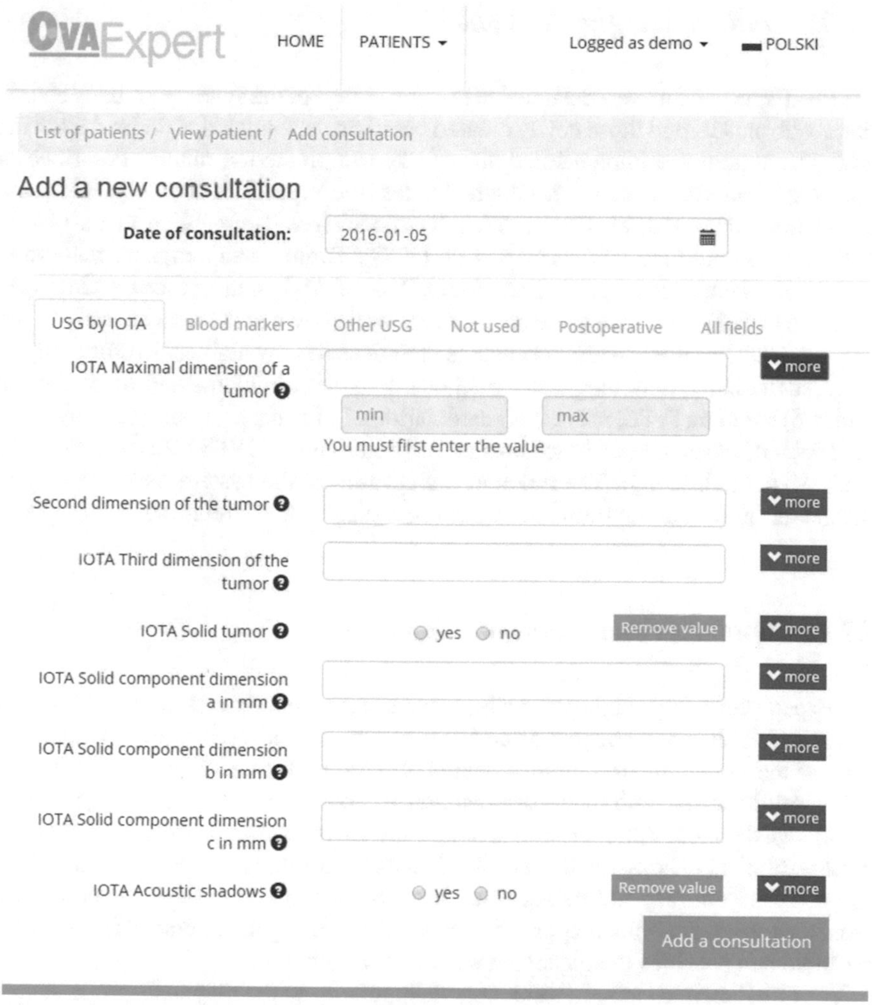

Fig. 5.1 An example of the OvaExpert user interface

of office, e.g. during consultation in the hospital, using a tablet or smartphone. A screenshot of the OvaExpert user interface is presented in Fig. 5.1.

At any time the doctor has access to the history of the patient's disease as well as the visualization of her diagnostic process. Throughout the diagnostic process, the system supports the physician by indicating which future diagnostic tests should be performed to improve the accuracy of the diagnosis. This solution is very helpful to the less experienced doctors. Moreover, it helps them protect patients from unnecessary tests and avoid increase in diagnostics costs.

5.1.3   Multiple Diagnostic Models

The OvaExpert system was designed to take advantage of the synergy of many classic diagnostic models and those newly created based on the knowledge derived from the data. The system has implemented all well-known prediction models described in Sect. 2.2.4, including Sonomorphological index (see Szpurek et al. [120]), Alcazar's model (see Alcazar et al. [2]), IOTA LR1 model (see Timmerman et al. [132]), IOTA LR2 model (see Timmerman et al. [132]), Timmerman's logistic regression model (see Timmerman et al. [129]) and Risk of Malignancy Index (RMI) (see Jacobs et al. [63]). Because these are methods well known to medical specialists and they trust their results, OvaExpert makes their decisions available, also for comparison. Additionally, new diagnostic methods have been implemented (described in Sect. 4.6) based on IVFS, using their cardinalities, IVFS aggregations (see Żywica et al. [166]) and classification based on similarity measures of IVFS (see Żywica [162], Stachowiak et al. [114]). The modular architecture of the system makes it easy to expand the system in the future to include new diagnostics modules.

5.1.4   Bipolar Presentation of Diagnosis

OvaExpert presents a diagnosis in a bipolar form (see Stachowiak et al. [113]). This means that it shows the degree of certainty for the diagnosis of tumor malignancy, while at the same time presenting a degree of certainty for the diagnosis of a benign tumor. Additionally, with such a presentation, it is possible to visualize the amount of knowledge deficiency and certainty of diagnosis. In a classic approach, the diagnostic process requires making the most probable diagnosis, although there are indications that this decision may not be accurate. It is obvious that in case of doubt about a given diagnosis, the bipolar approach—a positive and negative perspective—is very desirable and provides the physician with more information.

The OvaExpert system, in modeling the bipolarity in the diagnosis process, uses an approach based on Atanassov's intuitionistic fuzzy sets (IFSs) (see Atanassov [5], Wygralak [150]) described in more detail in Sect. 4.3. This approach is innovative in medicine and presents a diagnosis in a bipolar form. This is in line with the basic premise of the OvaExpert system which must be able to accept and process uncertain data. The patient's condition is described, on the one hand, by the degree at which the tumor is considered malignant and, on the other hand, by the degree to which it is considered benign. These two degrees do not have to sum up to 100% and the system may propose further tests to increase the reliability and completeness of the diagnosis (see Fig. 5.2).

5.1.5 Gathering of Expert Knowledge

The system allows users to enter their own diagnostic rules and thus create their own separate predictive modules (these modules do not affect the other predictive models implemented in the system and are to be used by only one particular physician). The diagnostic rules are in the form of fuzzy IF-THEN conditional sentences, where variables can be both numeric (such as marker level in blood) and linguistic fuzzy variables. With this solution, a particular physician can model his or her expertise reflecting his or her experience. At the same time, this can potentially lead to increased system diagnostic efficacy.

5.1.6 System Application in the Diagnosis Process

The diagnostic process begins with the medical history that the physician enters into the system. It is then continued through several steps at which further medical examinations are performed. Each of these steps will provide a better understanding of the patient's condition (e.g. blood markers, ultrasound examinations, etc.). The process is iterative. At each stage, OvaExpert computes bipolar diagnosis suggestion which consists of several possible diagnoses for a particular type of tumor, along with an indication of how likely it is that the diagnosis is correct and how likely it is that the diagnosis is incorrect (see Fig. 5.2). The system also recommends (if possible) a concrete diagnosis.

On the basis of this recommendation the doctor must decide whether further tests should be conducted or a final diagnosis should be made. The system supports the choice of an optimal diagnostic path. This is done by using knowledge from retrospective data (e.g. statistical methods) as well as using fuzzy rules introduced into the system by experts. The interaction with the physician is illustrated in Fig. 5.3.

5.1.7 System Architecture

A very important aspect in software development is selecting the right technology. The main factors that were taken into account when designing OvaExpert were stability, reliability, security, and high usability on every device, including smartphones. Additionally, the selected technology needed to meet modern software development standards. It was also important that the software was available under an open source license, which is supported by a large community of users.

OvaExpert was built using modern tools and technologies such as Java, Spring, AngularJS and PostgreSQL. The system uses the client-server approach, MVC pattern (Model-View-Controller) and is available through a web browser based on

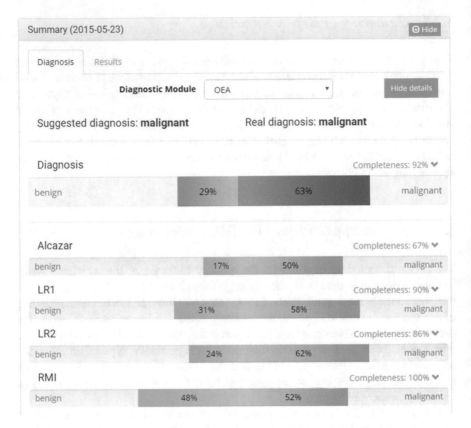

Fig. 5.2 A screenshot of the OvaExpert system presenting the recommended diagnosis in a bipolar form

RESTful web services. In the remaining part of this chapter the details of selected technologies and their role in the project will be discussed

The server of the OvaExpert system is based on Apache Tomcat (see The Apache Software Foundation [127]), which makes it possible to run a Java-based HTTP server. As the main framework we selected Spring (see Pivotal Software Inc. [99]), which was used to build the following key components of the application:

- Management of object life cycle and dependency with Spring IoC container;
- RESTful Web Services and MVC pattern using Spring MVC;
- security system using Spring Security;
- Spring Object-Relational Mapping (ORM).

For database management system (DBMS) we selected the open-source system PostgreSQL (see The PostgreSQL Global Development Group [128]), which is known for its reliability and ensures high data integration.

On the client side, the application was created with JavaScript and HTML using the AngularJS framework (see Project: AngularJS [105] and Bootstrap CSS (see Project:

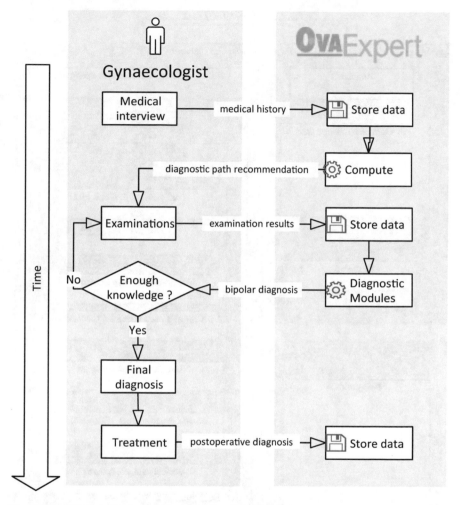

Fig. 5.3   A diagram of interaction between the physician and the OvaExpert system

Bootstrap [106]). The client and server communicate via RESTful web services and the JSON data format (see Internet Engineering Task Force (IETF) [59]). The client's application is available on the latest web browsers and is optimized for different screen sizes (computer, tablet, smartphone).

Additionally, the OvaExpert system uses several libraries to implement specific functions. Import and export to other data formats have been implemented using Apache POI library (see The Apache Software Foundation [126]) and OpenCSV (see OpenCSV Project Homepage [93]). Internationalization of the application was done using the *angular-translate* library (see Project: Aangular-translate [103]). Anonymization of medical data was performed with the *angular-cryptography* library (see Project: Angularcryptography [104]).

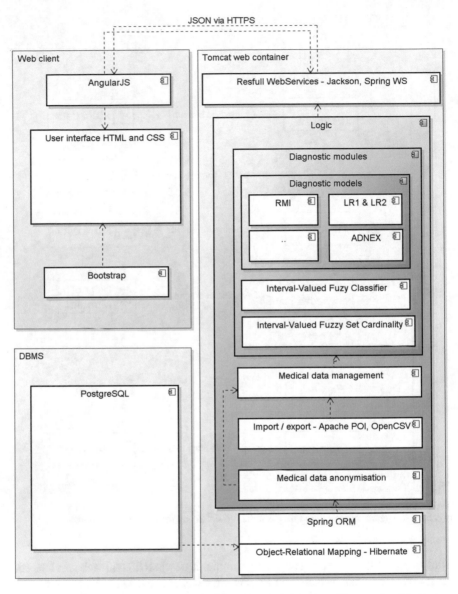

Fig. 5.4 UML diagram showing key features of the OvaExpert system (Żywica et al. [164])

The system architecture is presented in Fig. 5.4 which shows key elements of the system along with the technologies used to create them. Two elements deserve special attention due to their further use in medical practice. With medical data anonymization, it is easier for both the gynecologists and researchers to use confidential data in the system. Diagnostic modules fulfill the main goal of the OvaExpert project. The theoretical foundations for the construction of the diagnostic modules are discussed

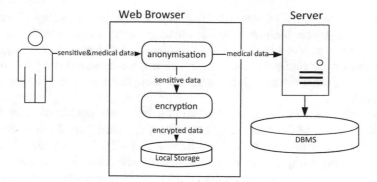

Fig. 5.5 Method of medical data anonymization in the OvaExpert system

in Chap. 2. The next section will deal with the technical aspects of medical data anonymization.

5.1.7.1 Anonymization of Medical Data

Personal information (especially medical data) is sensitive information. It is protected by law. In order to conduct research based on data collected by the system, the data must be anonymized. OvaExpert has been structured in such a way as to make the process automatic. Sensitive data are not sent to the server, and the physician has access to his patients' data.

Figure 5.5 shows the process of storing medical data. The doctor in the client application enters all medical data. Then the data are anonymized. Data are divided into two parts and a unique identifier is generated. The anonymized medical data are sent to the OvaExpert server, while personal information is encrypted and stored locally in the local web browser resource. This ensures that sensitive data remain on the doctor's computer and are only accessible to authorized users. On the other hand, scientists can use the collected anonymized medical data to create new methods and improve the existing ones. When a doctor needs to retrieve data from OvaExpert, sensitive information is downloaded and decrypted from the local browser resource, and connected with the anonymized medical data retrieved from the server (using a unique identifier).

5.2 Diagnostic Models

The main advantage of the system is its modular architecture. To provide the physician with the best possible diagnosis suggestion, it uses both the existing, recognized diagnostic methods as well as those newly created through data analysis and computational methods. All of them are integrated to improve the quality of the diagnosis. Currently, three high-level decision-making modules are implemented: *Based on IVFSs aggregation, based on IVFSs cardinality* and *based on prototypes and uncertainty-aware similarity measure*. In addition, the system is designed to allow the addition of other modules, including those using completely new computational techniques.

When creating the system, we decided not to build new models based on classic data classification methods such as SVM or Bayesian classifiers. Such models have already been investigated (see Van Holsbeke et al. [142]), but they are not used or even acknowledged by physicians for whom they are often too complex. For this reason, our methods are based on classifiers that are recognized by physicians in the field and we use fuzzy-based methodology because of its natural and accessible interpretation.

Various decision-making modules are available in the system only to physicians—experts who are involved in the process of creating the system. Because the system was built to support the less experienced physicians, there is only one, optimized decision model available to them.

The number of different diagnostic models available is quite large (see Sect. 2.2.4). In addition, there is no general agreement as to which one should be used in a particular case. Moreover, original models were not created to deal with missing data, even though incomplete information is a natural and commonplace phenomenon in medicine (see Hatch [52]). That is why it was a great challenge to support physicians in making effective diagnoses with imperfect information at their disposal.

5.2.1   Decision Module Based on Aggregation of Diagnostic Models (OEA)

One of the solutions proposed by the team is to use the synergy effect of many different diagnostic models and aggregate their results to achieve high quality decisions despite uncertain data.

Our previous research has shown that fuzzy aggregation methods prove very effective in improving the quality of diagnosis and minimizing the impact of lack of data and imprecision. This is due to the variety of models and their different levels of efficacy across different patient groups. Many models, when used simultaneously, considerably improve the quality of the decision.

As part of the research we developed the methodology of optimization, selection of aggregation functions and method of decision quality evaluation. During its application the most effective diagnostic method of aggregation was selected. The following aggregation methods were considered:

1. Weighted r-means,
2. Choquet and Sugeno integrals,
3. Triangular operations,
4. Ordered Weighted Average (OWA).

The methods listed above were considered both as operating on both numerical and interval-valued data.

The main approach deployed in the system is based on Ordered Weighted Averaging aggregation (OWA, see Yager [153]) and abbreviated as *OEA*. The algorithm based on this measure achieves efficacy far exceeding those achieved by single diagnostic models, despite data insufficiency. More details on this method can be found in Żywica et al. [166].

5.2.2 Decision Module Based on Interval-Valued Fuzzy Set Cardinality (FSC)

Another decision module implemented in the OvaExpert system is based on the method proposed in Chap. 4. This method is based, just like the previous method, on the synergy of known diagnostic models. What is different, however, is that it uses counting methods of IVFSs. It is highly effective in making diagnostic decisions and has very high decisiveness even in the case of considerable data insufficiency (see comparison of diagnostic effectiveness presented in Sect. 5.2.4).

5.2.3 Decision Module Based on Prototypes and Uncertainty-Aware Similarity Measure (IVFC)

This diagnostic module implements the innovative concept of Interval-Valued Fuzzy Classifier based on the uncertainty-aware similarity measure proposed and described in Stachowiak et al. [114] and generalized relative cardinality (see Żywica et al. [165]). A detailed description of the method can be found in Żywica [162], Żywica et al. [164].

This method, unlike the previous ones, allows us to assign tumors not only to one of two classes: malignant and benign (binary classification), but it also predicts the histopathological type of tumor and, thus, performs multi-class classification. The idea behind it is to keep all information about the data during the classification process, i.e. also information about the missing data. The classifier has been designed to deal with situations in which both the classified objects and classes themselves are imprecise, subjective and/or incomplete. In such cases, the obtained classifications are also uncertain.

The general idea is to prepare an interval-valued prototype vector for each class. Both the classes and data of specified patients are encoded using interval-valued fuzzy sets. They are compared using an interval-valued similarity measure that maintain uncertainty (see Stachowiak et al. [114]). This algorithm will be referred to as IVFC (OvaExpert decision method based on interval-valued fuzzy classifier).

5.2.4   Prognostic Efficacy of Individual Decision Modules

We will now compare the three predictive methods described above. The prognostic results of all three decision modules are presented in Table 5.1, and for the purpose of comparison the results for the original diagnostic models are also presented. Results obtained from the test data set are described in Sect. 4.7.

The original diagnostic models differ in their classification properties: some of them tend to make more conservative decisions (i.e. LR1, LR2, SM), and some of them are more liberal (i.e. RMI, Tim.). This can be observed in discernible differences in values between sensitivity and specificity. Only one of these models ensures the balance of both factors (Alc.). It should be noted that all original models have very low decisiveness (due to deficiencies in diagnostic data), which results in high total cost (calculated with cost matrix defined in Sect. 2.3.2.8).

The models implemented in the OvaExpert system have high sensitivity and specificity values. Two of them tend to be more conservative (*OEA* and *IVFC*), while *FSC* is more balanced. All three models provide a high level of decisiveness because they are able to deal with deficiencies in data. This is why their total cost is much lower than the original models.

A comparison of statistical significance between the original models and models offered in the OvaExpert system is presented in Table 5.2.

It can be noted that the diagnostic models of the OvaExpert system differ significantly from the original models in terms of classification. Although diagnostic

Table 5.1 Quality measures of the original diagnostic models compared with the new models of the OvaExpert system. Results obtained on test set

		Total cost	Dec. (%)	Sen. (%)	Spec. (%)	Acc.(%)
Original models	Alc. [2]	189.0	20.6	88.2	89.5	88.9
	LR1 [134]	184.0	27.4	92.6	57.1	77.1
	LR2 [134]	164.0	33.1	94.3	65.2	82.8
	RMI [63]	156.0	56.6	75.9	87.1	83.8
	SM [120]	142.0	62.9	94.6	71.2	79.1
	Tim. [129]	159.0	47.4	66.7	97.1	91.6
New diag. modules	OEA	72.0	96.6	90.2	86.4	87.6
	IVFC	72.5	100.0	90.4	84.6	86.3
	FSC	67.0	93.7	90.0	90.2	89.4

Table 5.2 The McNemar test with Benjamini-Hochberg correction between the original diagnostic models and the decision modules ($\alpha = 0.05$). Results obtained on test set

		Decision modules		
		OEA	IVFC	FSC
Original models	Alc. [2]	<0.001	<0.001	<0.001
	LR1 [134]	<0.001	<0.001	<0.001
	LR2 [134]	< 0.001	<0.001	<0.001
	RMI [63]	<0.001	<0.001	<0.001
	SM [120]	<0.001	<0.001	<0.001
	Tim. [129]	<0.001	<0.001	<0.001
Decision modules	OEA	–	0.579	0.579
	IVFC	0.579	–	0.093
	FSC	0.579	0.093	–

modules differ in classification quality indicators, the differences in classification are not statistically significant.

In light of these results, the OvaExpert system based on the presented modules is a promising tool for supporting the prognosis of ovarian tumors, especially in the case of partial gaps in diagnostic data that are common in the everyday medical practice.

Chapter 6
Summary

This monograph presents methods of supporting decision-making process in medical diagnosis based on IVFSs. Their undeniable advantage is that they can make effective decisions despite gaps in the input data. These methods were used in the implementation of the Intelligent Diagnostic System for Ovarian Tumors OvaExpert, which is also described in this publication.

The presented results of the efficacy of the described algorithms confirm their high usefulness. They allow effective integration of knowledge from many decision sources. The positive feedback from physicians testing the prototype version of the system holds promise as to the future development and application of the OvaExpert system in wider medical practice.

It is worth mentioning here the ongoing research concerning the OvaExpert project and future directions of its development.

We aim to make OvaExpert a future universal reference system for decision-making in the diagnosis of ovarian tumor.

We also hope that it will become a system for unified gathering medical data on ovarian cancer cases in Poland. This will contribute to further improvement of diagnostic methods and will consequently translate into a decrease in mortality among women.

6.1 Ongoing Research

The work on the OvaExpert system is still ongoing as the works and development opportunities are enormous. First, we will present research directions which have already yielded first promising results.

© Springer International Publishing AG 2018
K. Dyczkowski, *Intelligent Medical Decision Support System Based
on Imperfect Information*, Studies in Computational Intelligence 735,
https://doi.org/10.1007/978-3-319-67005-8_6

6.1.1 Using Fuzzy Control (Rule-Based System) as a Prediction Method

It was natural for us to decide to use a fuzzy decision-making system. This method has several advantages. First, it is easy to implement the very popular IOTA Simple Rules decision model (see Sect. 2.2.4.5). Second, it is possible to fuzzify it, thus allowing a better reflection of the perceptive character of the parameters entered into it. The last and perhaps most important advantage of the module is that an experienced physician can define his or her own decision-making methods based on his or her years long experience.

The implementation of such a model in the OvaExpert system and the first results of its validation were described in Baranowski [8] written as part of the OvaExpert project.

6.1.2 Application of Deep Learning

Another method that was of interest to our team was the use of deep learning methods to build a new prognostic model. Deep machine learning provides a new approach to neural network modeling and is becoming more and more popular. The condition for its application is to have a large enough set of learning data available. The first results were presented in Flieger [42]

6.1.3 Using Stochastic Orderings for Classifier Evaluation

An important element of our research is the methodology for selecting the optimal interval-valued classification methods. This aspect is explored more thoroughly in our further research. We were interested in non-classical methods of classification evaluation, using the generalized stochastic ordering proposed in Couso and Sánchez [19]. The first research results are promising and were published in Basiukajc [9] and also in Żywica et al. [163].

6.2 Future Plans for Development

A separate issue is the development of a method for proposing a personalized diagnostic path for a particular patient, depending on the outcome of subsequent diagnostic steps. The methods currently in use can indicate the next stage of diagnosis based on the minimization of hesitation margins. It is necessary, however, to broaden the study of this problem and to use other elements of knowledge.

We would also like to use the developed methods in the future in other medical fields, such as cardiac diagnosis. Our first insight into the topic makes us extremely optimistic about the research in this area.

References

1. Aamodt, A., Plaza, E.: Case-based reasoning: foundational issues, methodological variations, and system approaches. AI Commun. **7**(1), 39–59 (1994)
2. Alcazar, J.L., Merce, L.T., et al.: A new scoring system to differentiate benign from malignant adnexal masses. Obstet. Gynecol. Surv. **58**(7), 462–463 (2003)
3. Amor, F., Vaccaro, H., Alcázar, J.L., León, M., Craig, J.M., Martinez, J.: Gynecologic imaging reporting and data system a new proposal for classifying adnexal masses on the basis of sonographic findings. J. Ultrasound Med. **28**(3), 285–291 (2009)
4. Atanassov, K.T.: Intuitionistic fuzzy sets. Fuzzy Sets Syst. **20**(1), 87–96 (1986)
5. Atanassov, K.T.: Intuitionistic Fuzzy Sets. Springer, Heidelberg (1999)
6. Atanassov, K.T.: Intuitionistic fuzzy logics. Studies in Fuzziness and Soft Computing, vol. 351. Springer, Heidelberg (2017)
7. AxSys Technology Ltd: http://www.axsys.co.uk/ (2016). Accessed 18 July 2016
8. Baranowski, M.: Ovarian tumor diagnosis support with use of fuzzy controllers. Master's thesis, Adam Mickiewicz University in Poznań Poznań (2016). (in Polish)
9. Basiukajc, K.: Efectiveness analysis of machine learning algorithms for low quality data. Master's thesis, Adam Mickiewicz University in Poznań Poznań (2016). (in Polish)
10. Beliakov, G., Bustince, H., James, S., Calvo, T., Fernandez, J.: Aggregation for Atanassov's intuitionistic and interval valued fuzzy sets: the median operator. IEEE Trans. Fuzzy Syst. **20**(3), 487–498 (2012)
11. Belle, A., Kon, M.A., et al.: Biomedical informatics for computer-aided decision support systems: a survey. Scient. World J. **2013** (2013)
12. du Bois, A., Rochon, J., Pfisterer, J., Hoskins, W.J.: Variations in institutional infrastructure, physician specialization and experience, and outcome in ovarian cancer: a systematic review. Gynecol. Oncol. **112**(2), 422–436 (2009)
13. Bury, J., Hurt, C., Bateman, C., Atwal, S., Riddy, K., Fox, J., Saha, V.: Lisa a clinical information and decision support system for collaborative care in childhood acute lymphoblastic leukaemia. In: Proceedings of the AMIA, vol. 988 (2002)
14. Bury, J., Hurt, C., Roy, A., Cheesman, L., Bradburn, M., Cross, S., Fox, J., Saha, V.: Lisa: a web-based decision-support system for trial management of childhood acute lymphoblastic leukaemia. Br. J. Haematol. **129**(6), 746–754 (2005)
15. Bustince, H., Montero, J., Pagola, M., Barrenechea, E.: A survey of interval-valued fuzzy sets. In: Pedrycz, W., Skowron, A., Kreinovich, V. (eds.) Handbook of Granular Computing. Wiley, New Jersey (2008)
16. Chen, S.J., Hwang, C.L.: Fuzzy Multiple Attribute Decision Making Methods, pp. 289–486. Springer, Heidelberg (1992)
17. Clarity Informatics: Prodigy Homepage. http://prodigy.clarity.co.uk/ (2016). Accessed 18 July 2016

© Springer International Publishing AG 2018   113
K. Dyczkowski, *Intelligent Medical Decision Support System Based on Imperfect Information*, Studies in Computational Intelligence 735, https://doi.org/10.1007/978-3-319-67005-8

18. Cooper, G.M.: Elements of Human Cancer. Jones & Bartlett Learning (1992)
19. Couso, I., Sánchez, L.: Generalized stochastic orderings applied to the study of performance of machine learning algorithms for low quality data. In: José, M.A., Bustince, H., Reformat, M. (eds.) Proceedings of the 2015 Conference of the International Fuzzy Systems Association and the European Society for Fuzzy Logic and Technology, pp. 1534–1541. Atlantis Press (2015). (An optional note)
20. Czerniak, J.M., Apiecionek, L., Zarzycki, H., Ewald, D.: Proposed caeva simulation method for evacuation of people from a buildings on fire. Adv. Intell. Syst. Comput. **401**, 315–326 (2016a)
21. Czerniak, J.M., Dobrosielski, W.T., Apiecionek, L., Ewald, D., Paprzycki, M.: Practical application of ofn arithmetics in a crisis control center monitoring. In: Fidanova, S. (ed.) Recent Advances in Computational Optimization: Results of the Workshop on Computational Optimization WCO 2015, pp. 51–64. Springer International Publishing, Cham (2016b)
22. Czerniak, J.M., Zarzycki, H., Ewald, D.: AAO as a new strategy in modeling and simulation of constructional problems optimization. Simulation Modelling Practice and Theory (2017)
23. Davies, A.P., Jacobs, I., Woolas, R., Fish, A., Oram, D.: The adnexal mass: benign or malignant? evaluation of a risk of malignancy index. BJOG Int. J. Obst. Gynaecol. **100**(10), 927–931 (1993)
24. Davison, A.C., Hinkley, D.V.: Bootstrap methods and their application, vol. 1. Cambridge University Press (1997)
25. De Dombal, F., Leaper, D., Staniland, J.R., McCann, A., Horrocks, J.C.: Computer-aided diagnosis of acute abdominal pain. Br. Med. J. **2**(5804), 9–13 (1972)
26. De Luca, A., Termini, S.: On the convergence of entropy measures of the fuzzy set. Kybernetes **6**, 219–227 (1977)
27. De Luca, A., Termini, S.: On some algebraic aspect of the measures of fuzzines. Fuzzy Information and Decision Processes (1982)
28. Delgado, M., Sánchez, D., Amparo Vila, M.: Fuzzy cardinality based evaluation of quantified sentences. Int. J. Appr. Reason. **23**, 23–66 (2000)
29. Deschrijver, G., Cornelis, C., Kerre, E.E.: On the representation of intuitionistic fuzzy t-norms and t-conorms. IEEE Trans. Fuzzy Syst. **12**(1), 45–61 (2004)
30. Deschrijver, G., Král, P.: On the cardinalities of interval-valued fuzzy sets. Fuzzy Sets Syst. **158**(15), 1728–1750 (2007)
31. Dubois, D., Gottwald, S., Hajek, P., Kacprzyk, J., Prade, H.: Terminological difficulties in fuzzy set theory-the case of "intuitionistic fuzzy sets". Fuzzy Sets Syst. **156**(3), 485–491 (2005)
32. Dubois, D., Liu, W., Ma, J., Prade, H.: The basic principles of uncertain information fusion. an organised review of merging rules in different representation frameworks. Inf. Fusion **32**, 12–39 (2016)
33. Dubois, D., Prade, H.: Fuzzy cardinality and the modeling of imprecise quantification. Fuzzy Sets Syst. **8**, 43–61 (1985)
34. Dubois, D., Prade, H.: On the use of aggregation operations in information fusion processes. Fuzzy Sets Syst. **142**(1), 143–161 (2004)
35. Dubois, D., Prade, H.: An overview of the asymmetric bipolar representation of positive and negative information in possibility theory. Fuzzy Sets Syst. **160**(10), 1355–1366 (2009)
36. Dyczkowski, K., Wójtowicz, A., Żywica, P., Stachowiak, A., Moszyński, R., Szubert, S.: An intelligent system for computer-aided ovarian tumor diagnosis. In: Intelligent Systems' 2014, pp. 335–343. Springer (2015)
37. Eidam Diagnostics Corporation: http://www.eidam.com/ (2016). Accessed 18 July 2016
38. Eurostat: Causes of death statistics. http://ec.europa.eu/eurostat/statistics-explained/index.php/Causes_of_death_statistics (2016). Accessed 11 July 2016
39. Evidence-Based Medicine Working Group: Evidence-based medicine. a new approach to teaching the practice of medicine. Jama **268**(17), 2420 (1992)
40. Fawcett, T.: An introduction to roc analysis. Pattern Recognit. Lett. **27**(8), 861–874 (2006)

41. Ferlay, J., Soerjomataram, I., Ervik, M., Dikshit, R., Eser, S., Mathers, C., Rebelo, M., Parkin, D., Forman, D., Bray, F.: GLOBOCAN 2012 v1.0, Cancer Incidence and Mortality Worldwide: IARC CancerBase No. 11. http://globocan.iarc.fr (2014). Accessed 11 July 2016
42. Flieger, M.: Deep learning application in ovarian tumor malignancy classification. Master's thesis, Adam Mickiewicz University in Poznań, Poznań (2016). (in Polish)
43. Fodor, J.C., Roubens, M.: Fuzzy Preference Modelling and Multicriteria Decision Support, vol. 14. Springer Science & Business Media (2013)
44. GIDEON Informatics: http://www.gideononline.com/about/ (2016). Accessed 18 July 2016
45. Goff, B.A., Mandel, L.S., Drescher, C.W., Urban, N., Gough, S., Schurman, K.M., Patras, J., Mahony, B.S., Andersen, M.R.: Development of an ovarian cancer symptom index. Cancer **109**(2), 221–227 (2007)
46. Gottwald, S.: Many-valued logic and fuzzy set theory. In: Höhle, U., Rodabaugh, S.E. (eds.) Mathematics of Fuzzy Sets, Logic, Topology, and Measure Theory. The Handbooks of Fuzzy Sets Series, pp. 5–89. Kluwer Academic Publishers (1999)
47. Gottwald, S.: Fuzzy Sets and Fuzzy Logic: The Foundations of Application–from a Mathematical Point of View. Springer, Heidelberg (2013)
48. Gotwald, S.: A note on fuzzy cardinals. Kybernetika **16**, 156–158 (1980)
49. Grattan-Guinness, I.: Fuzzy membership mapped onto intervals and many-valued quantities. Math. Logic Quart. **22**(1), 149–160 (1976)
50. Halls, S.: BI-RADS Homepage. http://breast-cancer.ca/bi-rads/ (2016). Accessed 19 July 2016
51. Han, P., Klein, W., Arora, N.K.: Varieties of uncertainty in health care: a conceptual taxonomy. Med. Decis. Mak. **31**(6), 828–838 (2011)
52. Hatch, S.: Snowball in a Blizzard: A Physician's Notes on Uncertainty in Medicine. Basic Books, New York (2016)
53. Heintz, A., Odicino, F., Maisonneuve, P., Quinn, M., Benedet, J., Creasman, W., Ngan, H., Pecorelli, S., Beller, U.: Carcinoma of the ovary. Int. J. Gynecol. Obstet. **95**, S161–S192 (2006)
54. van Herk, E.: A diagnostic decision support system for general practice. Ph.D. thesis, Erasmus University Rotterdam (1994)
55. Herrera-Viedma, E., Cabrerizo, F.J., Kacprzyk, J., Pedrycz, W.: A review of soft consensus models in a fuzzy environment. Inf. Fusion **17**, 4–13 (2014). (Special Issue: Information fusion in consensus and decision making)
56. IBM Corporation: IBM Watson for Oncology. http://www.ibm.com/watson/watson-oncology.html (2016). Accessed 18 July 2016
57. iCAD Inc.: http://www.icadmed.com/ (2016). Accessed 18 July 2016
58. InferMed (Elsevier): https://www.elsevier.com/solutions/infermed (2016). Accessed 18 July 2016
59. Internet Engineering Task Force (IETF): The javascript object notation (JSON) data interchange formath (RFC 7159). https://tools.ietf.org/html/rfc7159 (2016). Accessed 12 Nov 2016
60. IOTA: Educational material. http://www.iotagroup.org/index.php/educational-material (2016a). Accessed 19 July 2016
61. IOTA: LR2 and simple rules predictive IOTA models for ovarian cancer application. http://www.iotagroup.org/index.php/software/2-uncategorised/42-lr2-and-simple-rules (2016b). Accessed 19 July 2016
62. IOTA: Webpage for the ADNEX risk model to diagnose ovarian cancer. http://www.iotagroup.org/adnexmodel/ (2016c). Accessed 19 July 2016
63. Jacobs, I., Oram, D., et al.: A risk of malignancy index incorporating ca 125, ultrasound and menopausal status for the accurate preoperative diagnosis of ovarian cancer. BJOG Int. J. Obstet. Gynaecol. **97**(10), 922–929 (1990)
64. Japkowicz, N., Shah, M.: Evaluating Learning Algorithms: A Classification Perspective. Cambridge University Press, New York, NY, USA (2011)

65. Kacprzyk, J.: Group decision-making with a fuzzy majority via linguistic quantifiers. Part I: a consensory-like pooling. Cybern. Syst. **16**(2–3), 119–129 (1985a)
66. Kacprzyk, J.: Group decision-making with a fuzzy majority via linguistic quantifiers. Part II: a competitive-like pooling. Cybern. Syst. **16**(2–3), 131–144 (1985b)
67. Kacprzyk, J.: Group decision making with a fuzzy linguistic majority. Fuzzy Sets Syst. **18**(2), 105–118 (1986)
68. Kaufmann, A.: Introduction a la Theorie des Saus-Ensambles Flous, vol. IV. Masson, Paris (1977)
69. Klement, E., Mesiar, R.E.P.: Triangular Norms. Kluwer Academic Publishers, Dordrecht (2000)
70. Kosary, C.L.: Cancer of the ovary. In: Ries, L.A.G., Young Jr, J.L., Keel, G.E., Eisner, M.P., Lin, Y.D., Horner, M.J.D. (eds.) Cancer survival among adults: US SEER program, 1988–2001, Patient and tumor characteristics SEER Survival Monograph Publication, pp. 133–144. National Cancer Institute (2007)
71. Kosiński, W., Prokopowicz, P., Ślęzak, D.: Ordered fuzzy numbers. Bull. Polish Acad. Sci. Ser. Sci. Math. **51**(3), 327–338 (2003)
72. Král, P.: An axiomatic approach to cardinalities of if sets. In: Computational Intelligence, Theory and Applications, pp. 681–691. Springer (2005)
73. Laboratory of Computer Science at the Massachusetts General Hospital: Using decision support to help explain clinical manifestations of disease. http://www.mghlcs.org/projects/dxplain (2016). Accessed 18 July 2016
74. Lin, Y., Kerre, E.: An overwiew of fuzzy quantifiers, part i interpretation. Fuzzy Sets Syst. **95**, 1–21 (1998)
75. Mamdani, E.H.: Application of fuzzy algorithms for control of simple dynamic plant. Proc. Instit. Electr. Eng. **121**(12), 1585–1588 (1974)
76. Maśliński, S., Ryżewski, J.: Patofizjologia: podręcznik dla studentów medycyny. Wydawnictwo Lekarskie PZWL (2007). (in Polish)
77. Mendel, J.M., John, R.B.: Type-2 fuzzy sets made simple. IEEE Trans. Fuzzy Syst. **10**(2), 117–127 (2002)
78. Mendel, J.M., John, R.I., Liu, F.: Interval type-2 fuzzy logic systems made simple. IEEE Trans. Fuzzy Syst. **14**(6), 808–821 (2006)
79. Menger, K.: Statistical metrics. Proc. Nat. Acad. Sci. USA **8**, 535–537 (1942)
80. Micromedex Solutions: http://micromedex.com/ (2016). Accessed 18 July 2016
81. Microsoft Corp.: Microsofr Azure Platform. https://azure.microsoft.com/ (2016). Accessed 12 Nov 2016
82. Moore, R.E.: Interval analysis, vol. 4. Prentice-Hall Englewood Cliffs (1966)
83. Moore, R.G., McMeekin, D.S., Brown, A.K., DiSilvestro, P., Miller, M.C., Allard, W.J., Gajewski, W., Kurman, R., Bast, R.C., Skates, S.J.: A novel multiple marker bioassay utilizing he4 and ca125 for the prediction of ovarian cancer in patients with a pelvic mass. Gynecol. Oncol. **112**(1), 40–46 (2009)
84. Morotti, M., Menada, M.V., Gillott, D.J., Venturini, P.L., Ferrero, S.: The preoperative diagnosis of borderline ovarian tumors: a review of current literature. Archiv. Gynecol. Obstet. **285**(4), 1103–1112 (2012)
85. Moszyński, R.: Diagnostic possibilities of classification of ovarian tumors based on: morphological and Doppler ultrasound examination, biochemical and angiogenesis markers assessment and mathematical prognostic models. Poznan University of Medical Sciences (2014). (in Polish)
86. Moszyński, R., Żywica, P., Wójtowicz, A., Szubert, S., Sajdak, S., Stachowiak, A., Dyczkowski, K., Wygralak, M., Szpurek, D.: Menopausal status strongly influences the utility of predictive models in differential diagnosis of ovarian tumors: an external validation of selected diagnostic tools. Ginekol. Pol. **85**(12), 892–899 (2014)
87. National Cancer Institute: SEER Stat Fact Sheets: Ovarian Cancer. http://seer.cancer.gov/statfacts/html/ovary.html (2016a). Accessed 11 July 2016

88. National Cancer Institute: Undersatnding cancer. http://www.cancer.gov/about-cancer/understanding/what-is-cancer (2016b). Accessed 19 July 2016
89. OECD.Stat: Health status. http://stats.oecd.org/index.aspx?DataSetCode=HEALTH_STAT (2016). Accessed 11 July 2016
90. Oniko, A., Bobrowski, L., et al.: HEPAR i HEPAR II-komputerowe systemy wspomagające diagnozowanie chorób watroby. In: Proceedings of the 12th Conference on Biocybernetics and Biomedical Engineering, pp. 1–5. Warsaw, Poland (2001)
91. OpenClinical: ERA. http://www.openclinical.org/aisp_era.html (2016a). Accessed 18 July 2016
92. OpenClinical: LISA. http://www.openclinical.org/aisp_lisa.html (2016b). Accessed 18 July 2016
93. OpenCSV Project Homepage. http://opencsv.sourceforge.net/ (2016). Accessed 12 Nov 2016
94. OvaExpert Project: OvaExpert R scripts on Github. http://ovaexpert.github.io/ovarian-tumor-aggregation (2016). Accessed 12 Nov 2016
95. OvaExpert Project Homepage. http://ovaexpert.pl/en/ (2016). Accessed 12 Nov 2016
96. Ovarian Cancer National Alliance: Statistics. http://www.ovariancancer.org/about/statistics/ (2016). Accessed 10 July 2016
97. Pankowska, A., Wygralak, M.: On hesitation degrees in IF-set theory. In: International Conference on Artificial Intelligence and Soft Computing, pp. 338–343. Springer (2004)
98. Pankowska, A., Wygralak, M.: General IF-sets with triangular norms and their applications to group decision making. Inf. Sci. **176**(18), 2713–2754 (2006)
99. Pivotal Software Inc.: Spring Homepage. https://spring.io/ (2016). Accessed 12 Nov 2016
100. Polish Society of Oncology: Biała Księga-Strategia zwalczania nowotworów złośliwych jajnika w Polsce. https://pto.med.pl/content/download/10997/124020/file/Biala%20Ksiega.pdf (2015a). (in Polish)
101. Polish Society of Oncology: Zielona księga-rak jajnika: zapobieganie, rozpoznawanie, leczenie. https://pto.med.pl/content/download/11000/124040/file/Zielona%20Ksiega.pdf (2015b). (in Polish)
102. Powers, D.M.: Evaluation: from precision, recall and f-measure to roc, informedness, markedness and correlation. J. Mach. Learn. Technol. **2**(1), 37–63 (2011)
103. Project: Aangular-translate. https://angular-translate.github.io/ (2016). Accessed 12 Nov 2016
104. Project: Angular-cryptography. https://github.com/middleout/angular-cryptography (2016). Accessed 12 Nov 2016
105. Project: AngularJS. https://www.angularjs.org/ (2016). Accessed 12 Nov 2016
106. Project: Bootstrap. http://getbootstrap.com/ (2016). Accessed 12 Nov 2016
107. Przelaskowski, A.: Computer-aided diagnosis based on medical imaging-challenges and development perspectives. Acta Bio-Optica et Informatica Medica. Inzynieria Biomedyczna **16**(3), 245–253 (2010). (in Polish)
108. R Core Team: R: A Language and Environment for Statistical Computing. R Foundation for Statistical Computing, Vienna, Austria (2014). http://www.R-project.org
109. Ralescu, D.: Cardinality, quantifiers, and the aggregation of fuzzy criteria. Fuzzy Sets Syst. **69**, 355–365 (1995)
110. Schweizer, B., Sklar, A.: Associative functions and statistcal triangle inequalities. Publ. Math. Debrecen **8**, 169–186 (1961)
111. Schweizer, B., Sklar, A.: Probabilistic Metric Spaces. North Holland Publishing Company, Amsterdam (1983)
112. Society of Gynecologic Oncology: New FIGO ovarian cancer staging guidelines. https://www.sgo.org/clinical-practice/guidelines/new-figo-ovarian-cancer-staging-guidelines/ (2014). Accessed 11 Oct 2016
113. Stachowiak, A., Dyczkowski, K., Wójtowicz, A., Żywica, P., Wygralak, M.: A bipolar view on medical diagnosis in ovaexpert system. In: Andreasen, T., Christiansen, H., Kacprzyk, J., Larsen, H., Pasi, G., Pivert, O., De Tre, G., Vila, M.A., Yazici, A., Zadrożny, S. (eds.) Flexible Query Answering Systems 2015, pp. 483–492. Springer (2016)

114. Stachowiak, A., Żywica, P., Dyczkowski, K., Wójtowicz, A.: An Interval-Valued Fuzzy Classifier Based on an Uncertainty-Aware Similarity Measure. In: Intelligent Systems' 2014, pp. 741–751. Springer (2015)
115. Steadman, I.: IBM's Watson is better at diagnosing cancer than human doctors. http://www.wired.co.uk/article/ibm-watson-medical-doctor (2013). Accessed 18 July 2016
116. Stukan, M.: Gynecologic oncology calculators. http://gin-onc-calculators.com/ (2016). Accessed 19 July 2016
117. Stukan, M., Dudziak, M., et al.: Usefulness of diagnostic indices comprising clinical, sonographic, and biomarker data for discriminating benign from malignant ovarian masses. J. Ultrasound Med. 34(2), 207–217 (2015)
118. Szmidt, E.: Distances and Similarities in Intuitionistic Fuzzy ets. Studies in Fuzziness and Soft Computing, vol. 307. Springer, Heidelberg (2014)
119. Szpurek, D.: Clinical value estimation of the ultrasound morphological and Doppler indices, CA 125 and TPS antigens in early differentiation of the ovarian tumors malignancy. Poznan University of Medical Sciences (2004). (In Polish)
120. Szpurek, D., Moszynski, R., Zietkowiak, W., Spaczynski, M., Sajdak, S.: An ultrasonographic morphological index for prediction of ovarian tumor malignancy. Eur. J. Gynaecol. Oncol. 26(1), 51–54 (2005)
121. Szpurek, D., Moszyński, R., Sajdak, S.: Clinical value of the ultrasound doppler index in determination of ovarian tumor malignancy. Eur. J. Gynaecol. Oncol. 25(4), 442–444 (2004)
122. Szubert, S.: Ocena roli wybranych czynników proangiogennych w neowaskularyzacji raka jajnika. Ph.D. thesis, Poznan University of Medical Sciences, Poznan (2015). (in Polish)
123. Szubert, S., Wójtowicz, A., Moszyński, R., Żywica, P., Dyczkowski, K., Stachowiak, A., Sajdak, S., Szpurek, D., Alcazar, J.L.: External validation of the IOTA ADNEX model performed by two independent gynecologic centers. Gynecol. Oncol. 142(3), 490–495 (2016)
124. Słowinski, R., Wilk, S., et al.: Communication and decision support in the environment of telemedical services. Zarządzanie i Technologie Informacyjne 3, 150–166 (2008). (in Polish)
125. Tadeusiewicz, R.: Informatyka Medyczna. Uniwersytet Marii Curie-Skłodowskiej w Lublinie, Instytut Informatyki (2011). (in Polish)
126. The Apache Software Foundation: Apache POI-the Java API for Microsoft Documents. https://poi.apache.org/ (2016a). Accessed 12 Nov 2016
127. The Apache Software Foundation: Apache Tomcat Homepage. http://tomcat.apache.org/ (2016b). Accessed 12 Nov 2016
128. The PostgreSQL Global Development Group: PostgreSQL Homepage. https://www.postgresql.org/ (2016). Accessed 12 Nov 2016
129. Timmerman, D., Bourne, T.H., Tailor, A., Collins, W.P., Verrelst, H., Vandenberghe, K., Vergote, I.: A comparison of methods for preoperative discrimination between malignant and benign adnexal masses: the development of a new logistic regression model. Am. J. Obstet. Gynecol. 181(1), 57–65 (1999a)
130. Timmerman, D., Schwärzler, P., Collins, W., Claerhout, F., Coenen, M., Amant, F., Vergote, I., Bourne, T.: Subjective assessment of adnexal masses with the use of ultrasonography: an analysis of interobserver variability and experience. Ultrasound Obstet. Gynecol. 13(1), 11–16 (1999b)
131. Timmerman, D., Testa, A.C., Bourne, T., Ameye, L., Jurkovic, D., Van Holsbeke, C., Paladini, D., Van Calster, B., Vergote, I., Van Huffel, S., et al.: Simple ultrasound-based rules for the diagnosis of ovarian cancer. Ultrasound Obstet. Gynecol. 31(6), 681–690 (2008)
132. Timmerman, D., Testa, A.C., Bourne, T., Ferrazzi, E., Ameye, L., Konstantinovic, M.L., Van Calster, B., Collins, W.P., Vergote, I., Van Huffel, S., et al.: Logistic regression model to distinguish between the benign and malignant adnexal mass before surgery: a multicenter study by the International Ovarian Tumor Analysis Group. J. Clin. Oncol. 23(34), 8794–8801 (2005)
133. Timmerman, D., Valentin, L., Bourne, T., Collins, W., Verrelst, H., Vergote, I.: Terms, definitions and measurements to describe the sonographic features of adnexal tumors: a consensus opinion from the International Ovarian Tumor Analysis (IOTA) Group. Ultrasound Obstet. Gynecol. 16(5), 500–505 (2000)

134. Timmerman, D., Van Calster, B., Testa, A.C., Guerriero, S., Fischerova, D., Lissoni, A., Van Holsbeke, C., Fruscio, R., Czekierdowski, A., Jurkovic, D., et al.: Ovarian cancer prediction in adnexal masses using ultrasound-based logistic regression models: a temporal and external validation study by the iota group. Ultrasound Obstet. Gynecol. **36**(2), 226–234 (2010)

135. Trąpcżński, P.: Raport Analityczny-analiza rynkowa dla metod diagnostycznych nowotworów jajników, w tym metody wspomagania decyzji w diagnostyce opartej na obliczeniach inteligentnych oraz metod diagnostycznych w zakresie oporności na chemioterapie. PPNT, Poznań (2015). (in Polish)

136. USARAD Holdings, Inc.: http://usarad.com/ (2016a). Accessed 18 July 2016

137. USARAD Holdings, Inc.: Radiology-on-Demand. http://usarad.com/teleradiology-technology.html (2016b). Accessed 18 July 2016

138. Valentin, L.: Pattern recognition of pelvic masses by gray-scale ultrasound imaging: the contribution of doppler ultrasound. Ultrasound Obstet. Gynecol. **14**(5), 338–347 (1999)

139. Valentin, L., Ameye, L., Jurkovic, D., Metzger, U., Lécuru, F., Van Huffel, S., Timmerman, D.: Which extrauterine pelvic masses are difficult to correctly classify as benign or malignant on the basis of ultrasound findings and is there a way of making a correct diagnosis? Ultrasound Obstet. Gynecol. **27**(4), 438–444 (2006)

140. Van Calster, B., Van Belle, V., Vergouwe, Y., Timmerman, D., Van Huffel, S., Steyerberg, E.W.: Extending the c-statistic to nominal polytomous outcomes: the polytomous discrimination index. Statist. Med. **31**(23), 2610–2626 (2012)

141. Van Calster, B., Van Hoorde, K., Valentin, L., Testa, A.C., Fischerova, D., Van Holsbeke, C., Savelli, L., Franchi, D., Epstein, E., Kaijser, J., et al.: Evaluating the risk of ovarian cancer before surgery using the adnex model to differentiate between benign, borderline, early and advanced stage invasive, and secondary metastatic tumours: prospective multicentre diagnostic study. BMJ **349**, g5920 (2014)

142. Van Holsbeke, C., Van Calster, B., et al.: External validation of mathematical models to distinguish between benign and malignant adnexal tumors: a multicenter study by the International Ovarian Tumor Analysis Group. Clin. Cancer Res. **13**(15), 4440–4447 (2007)

143. Walker, C.L., Walker, E.A.: The algebra of fuzzy truth values. Fuzzy Sets Syst. **149**(2), 309–347 (2005)

144. World Cancer Research Fund International: Ovarian cancer statistics. http://www.wcrf.org/int/cancer-facts-figures/data-specific-cancers/ovarian-cancer-statistics (2016). Accessed 11 July 2016

145. World Health Organization: Mortality database. http://www.who.int/healthinfo/mortality_data/. Accessed 16 May 2016

146. Wygralak, M.: Fuzzy cardinals based on the generalized equality of fuzzy subsets. Fuzzy Sets Syst. **18**, 143–158 (1986)

147. Wygralak, M.: Triangular operations, negations, and scalar cardinality of fuzzy set. In: Zadeh, L.A., Kacprzyk, J. (eds.) Computing with Words in Information/Intelligent Systems 1. Foundations, pp. 1:326–341. Springer, Heidelberg (1999)

148. Wygralak, M.: Fuzzy sets with triangular norms and their cardinality theory. Fuzzy Sets Syst. **124**, 1–24 (2001)

149. Wygralak, M.: Cardinalities of Fuzzy Sets. Springer, Heidelberg (2003)

150. Wygralak, M.: Intelligent Counting Under Information Imprecision. Studies in Fuzziness and Soft Computing, vol. 292. Springer, Heidelberg (2013)

151. Wójtowicz, A., Żywica, P., Stachowiak, A., Dyczkowski, K.: Solving the problem of incomplete data in medical diagnosis via interval modeling. Appl. Soft Comput. **47**, 424–437 (2016)

152. Wójtowicz, A., Żywica, P., Szarzyński, K., Moszyński, R., Szubert, S., Dyczkowski, K., Stachowiak, A., Szpurek, D., Wygralak, M.: Dealing with uncertainty in ovarian tumor diagnosis. In: Krassimir Atanassov, E.A. (ed.) New Developments in Fuzzy Sets, Intuitionistic Fuzzy Sets. Generalized Nets and Related Topics. IBS PAN-SRI PAS, Warsaw (2014)

153. Yager, R.R.: On ordered weighted averaging aggregation operators in multicriteria decision making. IEEE Trans. Syst. Man Cybern. **18**(1), 183–190 (1988)

154. Zadeh, L.A.: Fuzzy sets. Inf. Control **8**(3), 338–353 (1965)

155. Zadeh, L.A.: The concept of a linguistic variable and its application to approximate reasoning-i. Inf. Sci. **8**(3), 199–249 (1975)
156. Zadeh, L.A.: A theory of approximate reasoning. In: Hayes, J.E., Michel, D., Mikulich, L. (eds.) Machine Inteligence, vol. 9, pp. 149–194. Willey, New York (1979)
157. Zadeh, L.A.: Fuzzy probabilities and their role in decision analysis. In: Proceedings of the IFAC Symposium on Theory and Applications of Digital Control, pp. 15–23. New Dehli (1982)
158. Zadeh, L.A.: A computational approach to fuzzy quantifiers in natural languages. Comput. Math. Appl. **9**(1), 149–184 (1983)
159. Zadeh, L.A.: Fuzzy logic = Computing with words. In: Zadeh, L.A., Kacprzyk, J. (eds.) Computing with Words in Information/Intelligent Systems 1. Foundations, pp. 3–23. Springer, Heidelberg, New York (1999)
160. Zadeh, L.A.: Toward extended fuzzy logic-a first step. Fuzzy Sets Syst. **160**(21), 3175–3181 (2009)
161. ZynxHealth: http://www.zynxhealth.com/ (2016). Accessed 18 July 2016
162. Żywica, P.: Similarity measures of interval-valued fuzzy sets in classification of uncertain data. applications in ovarian tumor diagnosis. Ph.D. thesis, Adam Mickiewicz University in Poznań (2016). (In Polish)
163. Żywica, P., Basiukajc, K., Couso, I.: Practical notes on applying generalised stochastic orderings to the study of performance of classification algorithms for low quality data. In: EUSFLAT 2017. Warsaw, Poland (2017). (To appear)
164. Żywica, P., Dyczkowski, K., Wójtowicz, A., Stachowiak, A., Szubert, S., Moszyński, R.: Development of a fuzzy-driven system for ovarian tumor diagnosis. Biocybern. Biomed. Eng. **36**(4), 632–643 (2016a)
165. Żywica, P., Stachowiak, A., Wygralak, M.: An algorithmic study of relative cardinalities for interval-valued fuzzy sets. Fuzzy Sets Syst. **294**, 105–124 (2016b)
166. Żywica, P., Wójtowicz, A., Stachowiak, A., Dyczkowski, K.: Improving medical decisions under incomplete data using interval-valued fuzzy aggregation. In: Proceedings of 9th European Society for Fuzzy Logic and Technology (EUSFLAT), pp. 577–584. Gijón, Spain (2015)

Index

Printed in the United States
By Bookmasters